MCQ's in PLANT BREEDING, BIOTECHNOLOGY & SEED SCIENCE

Related Books

• Basic Concepts of Plant Biotechnology (With MCQs)	*Vijay Prakash*
• Basic Concepts of Plant Science	*S.K. Bangarwa*
• Botany for NEET and other Medical Entrance **	*Khwaja Salahuddin*
• Competitive Nematology **	*N.G. Ravichandra*
• CSIR NET: Part A (Hindi)	*Christy Varghese*
• CSIR-NET General Aptitude: A New Outlook	*Christy Varghese*
• Forestry: A Subjective Guide for IFS Aspirants	*K.T. Parthiban*
• Fundamentals of Agriculture	*R.L. Arya*
• Glimpse on General Agriculture	*P. Laxman Rao*
• Indiras Objective Agricultural Biotechnology *2nd Ed*	*R.L. Arya*
• Indiras Objective Agricultural Extension	*R.L. Arya*
• Indiras Objective Agriculture	*R.L. Arya*
• Indiras Objective Agronomy *2nd Ed*	*R.L. Arya*
• Krishi Jagat: JET Evam Krishi Paryavekshak Pariksha Upoagi	*M.L. Jat*
• MCQ in Agricultural Microbiology**	*D. G.Panpatt*
• MCQs for Food Technology **	*P.M. Kotecha*
• Objective Agribusiness Management *3rd Ed*	*S.R. Panigrahy*
• Objective Animal Physiology for JRF, SRF, ARS, NET, Civil Services & Other Competitive Examinations	*Arpita Mohapatra*
• Objective Environmental Studies and Disaster Management	*M. Prasanthrajan*
• Objective Food Science & Technology *3rd Ed*	*Deepak Mudgil*
• Objective Forestry: For All Competitive Examination *3rd Ed***	*K.T. Parthiban*
• Objective Home Science at a Glance	*Shanti Balda*
• Objective Life Science MCQs for Life Science Examination *3rd Ed*	*Kailash Choudhary*
• Objective Plant Pathology *2nd Ed*	*Susanta Banik*
• Objective Plant Physiology *3rd Ed*	*P. Dwivedi*
• Objective Plant Science at a Glance	*D.K. Verma*
• Objective Seed Science and Technology	*K. Vanangamudi*
• Objective Type Questions and Answers in Veterinary Immunology	*T.R. Kannaki*
• Plant Pathology at a Glance (Encyclopedia of Plant Pathology)	*D.P. Tripathi*
• Question Bank: Seed Science and Technology	*K. Vanangamudi*
• Questions and Answers in Environmental Science Practical	*R.K. Kamble*
• Universal Forestry for IFS, JRF, SRF, NET, ACF, RFO, AFO, IBPS and Other Allied Exams *2nd Ed***	*Mohit Husain*
• Universal Objective Forestry for IFS, JRF, SRF, NET, ACF, RFO, AFO, IBPS and Other Allied Exams *2nd Ed***	*Mohit Husain*
• Vedic Maths and Mental Arithmetic: A New Outlook **	*Christy Varghese*
• Vegetable Crops at a Glance	*S.K. Tyagi*

For Full Content & Sample Page Visit Website

To know your local book store call or us on +91-95211-31111

MCQ's in
PLANT BREEDING, BIOTECHNOLOGY & SEED SCIENCE

K. Vanangamudi

Former Dean (Agriculture)
Former Professor & Head (Seed Science & Technology)
Tamil Nadu Agricultural University
Coimbatore 640 003, Tamil Nadu

R. Vinoth

Teaching Assistant (Plant Breeding & Genetics)
Institute of Agriculture,
Tamil Nadu Agricultural University,
Kumulur 621 712, Trichy, Tamil Nadu.

S. Elayabalan

Assistant Professor (Agricultural Biotechnology),
Imayam Institute of Agriculture and Technology,
Thuraiyur, Trichy 621 206, Tamil Nadu.

K . Raja

Associate Professor,
Department of Seed Science & Technology,
Tamil Nadu Agricultural University,
Coimbatore 640 003, Tamil Nadu

SCIENTIFIC PUBLISHERS

Published by

SCIENTIFIC PUBLISHERS (INDIA)
5 A, New Pali Road, P.O. Box 91
Jodhpur 342 001, INDIA
E-mail: info@scientificpub.com
Website: http://www.scientificpub.com

ISBN : 978-93-88449-30-4
eISBN : 978-93-88449-31-1

Printed in India

Preface

Plant breeding is an important field in agriculture which deals with the development of plants with desired traits. This area of science helps for ensuring the food security by developing new varieties with higher yield and quality, pest and disease resistance, drought tolerance, etc. It is practiced throughout the world by individuals such as Plant Breeders, Farmers, Gardeners etc.

Biotechnology includes the technological applications in the biological system to make or modify the products for specific use. Medical, agricultural, industrial and environmental biotechnologies are the key areas of this field. It is the fast growing field in the recent science with the adoption of genetic engineering technologies.

Seed science is the study of the seed biology and seed production. Generally, seed is an important input in agriculture and any agriculture begins and ends with the seed. Seed biology involves several sciences beginning from syngamy until it is sown in the field. Also, many organizations are involved in the production and quality control system of the crop seeds throughout the world.

Considering the importance of Plant breeding, Agricultural Biotechnology and Seed Science, it is contemplated to prepare the book on "MCQ's in Plant Breeding, Agricultural Biotechnology and Seed Science". The main aim is to expose the students those who are preparing for the competitive examinations like Agricultural Research Services, Public Service Commissions, Institute of Banking Personnel Selection, University and Institute admissions etc.

The book has been prepared consisting of vast questions with the multiple choice answers in three major parts. Evidently, about 9 chapters with 600 questions in Plant Breeding, 8 chapters with 700 questions in Agricultural Biotechnology and 63 chapters with 2000 questions in Seed Science and Technology are compiled in one consortium. Genetics, breeding methods of self pollinated, cross pollinated and asexually propagated crops, resistance breeding, mutation breeding and polyploidy breeding in Plant Breeding; cell biology, molecular biology, tissue culture, animal biotechnology and bioinformatics in Agricultural Biotechnology; and seed formation, biology, production, post harvest processing, storage, health, marketing and legislation in Seed Science and Technology are some of the important chapters covered in the book.

Since this book is prepared with latest information, we are sure that the book will be insightful to the Scientists and Students for updating their knowledge and also who are preparing the examinations.

Authors

Contents

Chapter 1

GENETICS

1. Cytogeneticaly, the definitive nucleus is :
(A) Haploid (B) Diploid
(C) Triploid (D) Tetraploid

2. Crossing over occurs at :
(A) Two strand stage
(B) Three strand stage
(C) Four strand stage
(D) None of the above

3. Chromosome numbers in somatic cells of a maize plant are :
(A) 28 (B) 34
(C) 7 (D) 20

4. Chromosome which pair at meiotic prophase and are similar size, shape, structure and function are known as :
(A) Isochromosome (B) Homologus
(C) Isocentric (D) Acentric

5. Clone in general is :
(A) Heterogenous (B) Homozygous
(C) Heterozygous (D) Homomorphy

6. Commercial banana have ------------ instead of 22 chromosomes of ordinary diploid varieties.
(A) 13 (B) 23
(C) 33 (D) 43

7. Complementary gene produces phenotypic ratio :
(A) 9:7 (B) 9:6:1
(C) 15:1 (D) 13:3

8. Exchange of genetic material takes place between non-sister chromatids of homologus chromosomes during the :
(A) Two strand stage in prophase I
(B) Two strand stage in prophase II
(C) Metaphase I stage in meiosis
(D) Four strand stage in prophase I

9. Epistasis should be carefully distinguished from dominance, which refers to non-additivity of alleles at :
(A) The same locus of non-homologus chromosomes
(B) Different loci of non-homologus chromosomes
(C) The same locus of a homologus chromosome
(D) Different loci of a homologus chromosome

10. In wheat and oat there are ------------ genes for red versus white kernels.
(A) 2 (B) 3
(C) 4 (D) 5

11. Coefficient of variation is useful for comparing :
(A) Same population
(B) Different population
(C) Same treatments
(D) Value

12. A polyploid having chromosome sets from different sources such as different species is known as :
(A) Autopolyploid (B) Allopolyploid
(C) Tetraploid (D) Aneuploid

13. Nucleolus plays an important role in the production of :
(A) mRNA (B) tRNA
(C) rRNA (D) sRNA

14. A tautomeric shift in adenine (amino to imino) allows it to pair with ------------- only.
(A) Adenine (B) Guanine
(C) Cytosine (D) Thymine

15. The resulting two daughter cells produced by bridge-breakage fusion cycle will have :
(A) Deficiency
(B) Duplication
(C) Both (A) and (B)
(D) Inversion

16. Maximum absorbance of DNA molecule is at ------------- O_A wave length.
(A) 260 (B) 220
(C) 400 (D) 630

17. Okazaki fragments are formed due to synthesis of DNA in direction :
(A) 5'-3' (B) 3'-5'
(C) 7'-3' (D) 3'-7'

18. Harshey and Chase in their experiments labelled DNA with :
(A) 3H (B) ^{35}S
(C) ^{32}P (D) ^{35}T

19. The cell walls of fungi are made up of :
(A) Cellulose (B) Chitin
(C) Pectin (D) Proteins and lipids

20. A common crop which is grown of natural autopolyploid :
(A) Tomato (B) Rice
(C) Potato (D) Cotton

21. Dee - Gee - Woo - Gen is the source of dwarfing gene in :
(A) Ragi (B) Bajra
(C) Peanut (D) Rice

22. In pulses, limiting amino acids are :
(A) Methionine (B) Tryptophan
(C) Lusine (D) Both (A) and (B)

23. Wobble hypothesis was given by :
(A) Watson (B) Crick
(C) Benzer (D) Flor

24. Transposable elements were discovered by:
(A) McClintock (B) Shull
(C) Robert (D) Mendel

25. In C_4 pathway CO_2 combines with :
(A) Ribulose monophosphate
(B) Ribulose diphosphate
(C) Phosphoglyceric acid
(D) Phosphoenol pyruvate

26. The cytological identification of chromosomes depends mainly on?
(A) Length of chromosome and position of centromere
(B) Amount of DNA and intensity of staining
(C) Number of nucleoli and centromeres
(D) Number of chromatids and length of arms

27. In Oogenesis, the cell that corresponds to a spermatid is called a (an) :
(A) Ovum
(B) Egg
(C) Secondary oocyte
(D) Secondary polar body

28. If the cells of a plant root tip contain 16 chromosomes, how many chromosomes will be found in its pollen grains?
(A) 4 (B) 8
(C) 12 (D) 32

29. Nutritional improvement for the tryptophan content has been achieved in a ----------- cell-line.

(A) Tomato (B) Cucurbits
(C) Coffee (D) Tobacco

30. Observation of Hugo de Vries were based on :

(A) *Pisum sativum*
(B) *Drosophila melanogastor*
(C) *Solanum tuberosum*
(D) *Oenothera lamarkiana*

31. Operon concept was given by :

(A) Mosselson and Stahl
(B) Beedle and Tatum
(C) Watson and Crick
(D) Jacob and Manod

32. Paramutation in ------------ is perhaps the one and only example of blending of alleles.

(A) Sorghum (B) Maize
(C) Pearl millet (D) Foxtail millet

33. Pink (NN) flowered plants crossed with blue (nn) flowered plants gave black (Nn) flowered plants in F_1 generation. If F_1 plants are selfed, which of the following will be the phenotype of F_2 progeny?

(A) All blue
(B) 1/2 blue and 1/2 pink
(C) All pink
(D) 1/4 pink, 1/2 black, 1/4 blue

34. Plants grown from the seed of an open-pollinated variety of maize :

(A) Vary in plant height but are uniform in seed characters
(B) Are uniform in plant height and seed characters
(C) Vary in plant height as well as seed characters
(D) Are uniform in plant height but vary in seed characters

35. Replacement of purine by pyrimidine and *vice-versa* is known as :

(A) Translocation (B) Transduction
(C) Transversion (D) All the above

36. The genetic material in TMV is :

(A) Protein (B) DNA
(C) RNA (D) Carbohydrate

37. Exchange of segments between two non-homologous chromosomes is known as :

(A) Crossing over
(B) Reciprocal translocation
(C) Inversion
(D) Duplication

38. Sex in *Drosophila* is governed by the following mechanism :

(A) OX method (B) ZZ method
(C) XY method (D) X autosomes ratio

39. Single gene affecting more than one character is :

(A) Polymorphism
(B) Phenocopy
(C) Pleiotropy
(D) Complete penetrance

40. The application of principles of genetics for the improvement of mankind is :

(A) Linkage
(B) Eugenics
(C) Population genetics
(D) Heredity

41. ICRISAT in Hydrabad was established during :

(A) 1961 (B) 1907
(C) 1972 (D) 1955

42. The "Sonalika" variety of wheat has dwarfing genes derived from :

(A) Norin - 10 (B) Rht 1 - 10
(C) Norin - 1 (D) Stu 1

43. Tetraploid potato has 48 chromosomes, whereas ordinary diploid species have ------------ chromosomes.

(A) 34 (B) 16
(C) 14 (D) 24

44. System of classification proposed by Hutchinson is :

(A) Natural (B) Artificial
(C) Numerical (D) Phylogenetic

45. Superiority of heterozygote leads to :

(A) Fixation of recessive allele
(B) Fixation of dominant allele
(C) Maintenance of both the alleles
(D) Depression of inbreeding

46. Sunflower is :

(A) Annual herb (B) Biennial herb
(C) Perennial herb (D) None of the above

47. Failure of paring of homologous chromosome in prophase of meiosis is known as :

(A) Disjunction (B) Asynapsis
(C) Synapsis (D) Univalent shift

48. If a dihybrid is test crossed, the phenotypic ratio of progeny will be :

(A) 9 : 3 : 3 : 1 (B) 9 : 3 : 4
(C) 1 : 1 : 1 : 1 (D) 15 : 1

49. The ploidy level in generative nucleus is :

(A) n (B) 2n
(C) 3n (D) n + n

50. If a character is cytoplasmically controlled, then :

(A) The progeny resembles male parent
(B) The progeny resembles female parent
(C) 15:1 segregation in F_1
(D) A segregation of 3:1 is observed in F_1

51. ICRISAT deals with :

(A) Millets (B) Cotton
(C) Maize (D) Wheat

52. The proportion of individuals expression a particular gene combination in the phenotypes in a population is known as :

(A) Pedigree (B) Penetrance
(C) Expressivity (D) Phenogroup

53. The gene which is suppressed by another non-allelic gene in an interaction of genes is known as :

(A) Epistatic (B) Hypostatic
(C) Incomplete (D) Penetrance

54. How many cells will be produced from a cell if the cell divides meiotically?

(A) 1 (B) 2
(C) 3 (D) 4

55. ATP synthesis occurs in :

(A) Chloroplast
(B) Mitochondria
(C) Chloroplast, Mitochondria and Nuclei
(D) Chloroplast and Mitochondria

56. An organism with one extra chromosome (2n + 1), is instead of the normal diploid (2n) is known as :

(A) Monoploid (B) Trisomic
(C) Monosomic (D) Tetrasomic

57. Segregation can occur :

(A) In either of the nuclear division
(B) In one of the two nuclear division
(C) In the first nuclear division
(D) In the second nuclear division

58. Which of the following is true?

(A) Mitochondria is important constituents of plant and animal cells
(B) Mendelian inheritance is not applicable to qualitative traits only
(C) A variety resistance to a disease become tolerant to it due to the production of new races of the disease
(D) Mitochondria is not an important constituents of plant and animal cells

59. An organism having the gametic chromosome numbers is called as :

(A) Gamete (B) Genome
(C) Haploid (D) Hybrid

60. The total genetic information possessed by reproductive numbers of a population of sexually reproducing organism :

(A) Genetic equilibrium
(B) Gene bank
(C) Gene pool
(D) Acclimatization

61. Resemblance among individuals related by descent :

(A) Ancestry (B) Attavism
(C) Pedigree (D) Heredity

62. After the cell has undergone mitosis, the amount of DNA in daughter cells will be :

(A) Equal to the parent cell
(B) Double of the parent cell
(C) Half of the parent cell
(D) None of the above

63. Heritability may be defined as the :

(A) Sum total of hereditary material present in a species
(B) Interaction product of genotype with environment
(C) Proportion of phenotypic variability which is due to heredity
(D) Degree of resemblance between the original and the selected plants

64. The occurrence of differences among individuals :

(A) Evolution (B) Variation
(C) Heritability (D) All the above

65. Group of plants with similar characteristics of any rank is termed :

(A) Genus (B) Species
(C) Taxon (D) Order

66. Mendel is famous for :

(A) The discovery of penicillin
(B) The theory of mutation
(C) Laws of limiting factors
(D) Propounding the laws of inheritance

67. Average performance of a strain in a series of cross combination :

(A) Stability
(B) Competitive ability
(C) General combining ability
(D) Specific combining ability

68. Variation in phenotype among individuals of a species may be due to :

(A) Only genetical
(B) Only environmental
(C) General or environmental differences but not both
(D) Genetical or environmental differences or both

69. Given that the somatic chromosome number of *Triticum aestivum* is $2n=6x=42$, which one of the following pairs is correctly matched?

(A) Monosome ($2n = 40$)
(B) Trisome ($2n = 42$)
(C) Tetrasome ($2n = 44$)
(D) Nullisome ($2n = 41$)

70. Germplasm means :

(A) A breeding method
(B) A new pest
(C) A collection of genetic material
(D) A new disease

71. Genotype x Environment interaction would take place when there is/are :

(A) One genotype and two environments
(B) Two genotypes and one environment
(C) Three genotypes and one environment
(D) Two or more genotypes and two or more environments

72. A plant breeder is interested to control pollination to :
(A) Prevent cross pollination
(B) Control fertilization
(C) Allow cross pollination
(D) Control double fertilization

73. Flower colour in plant is a :
(A) Qualitative character
(B) Quantitative character
(C) Not a qualitative character
(D) Not a quantitative character

74. For Hardy-Weinberg law to operate :
(A) There should be control pollination
(B) There should be a selection
(C) The population should be large
(D) There should be pedigree selection

75. A mendelian population means :
(A) Inbred population
(B) Random-mating population
(C) Self-pollinated population
(D) Hybrid population

76. Objectives of population improvement :
(A) To maintain a homozygous balance
(B) Creates narrow genetic base
(C) To keep recombination in a limit
(D) Recombination and recovery of transgressive segregants

77. *Cannovalia gladiata* is a :
(A) Sword bean (B) Broad bean
(C) Lima bean (D) Butter bean

78. Which one of the following sets is the correct sequence of events in the cell?
(A) Transcription, translation, protein synthesis
(B) Transcription, protein synthesis, translation
(C) Translation, protein synthesis, transcription
(D) Protein synthesis, transcription, translation

79. Dr. Hargovind Khurana was awarded the Nobel Prize in 1968 for his work on :
(A) Mutation
(B) RNA synthesis
(C) Protein synthesis
(D) Genetic code and *in vitro* synthesis of DNA

80. The 'operon concept' of gene regulation was proposed by :
(A) Beadle and Tatum
(B) Jacob and Monod
(C) Watson and Crick
(D) None of these

81. Out breeding crops are mostly :
(A) Annuals (B) Perennials
(C) Bisexual (D) Decidious

82. Normally used test to evolve GCA in forage crops :
(A) RRS (B) RSGCA
(C) RSSCA (D) Poly cross test

83. Progeny of a single plant obtained by asexual reproduction is :
(A) Pure line (B) Inbred
(C) Clone (D) Strain

84. Chromosomes other than sex chromosomes are called
(A) Autosomes (B) Allosomes
(C) Autoallosomes (D) Heterosomes

85. An exchange of genetic material between homologous chromosomes is called :
(A) Recombination
(B) Mutation
(C) Introgression (D) Crossing over

86. Coupling and repulsion hypothesis was given by :
(A) Darlington (B) Bateson
(C) Morgan (D) Muller

87. Crop that is probably more diversified genetically than any other crop plant :

(A) Ragi (B) Sorghum
(C) Rice (D) Maize

88. Theory of 'Gemmules' was proposed by :

(A) Kolreuter (B) Lamark
(C) Darwin (D) Weisman

89. Incompatibility is very common in plant kingdom was reported by :

(A) Lewis (B) Mather
(C) East (D) None of these

90. D-genome of common cultivated wheat has been contributed by :

(A) *T. monococcum*
(B) *T. durum*
(C) *T. timophivi*
(D) *Ae squarosa*

91. In a completely linked condition, the crossover frequency will be :

(A) Nil (B) <25
(C) <75 (D) <100

92. Homozygous recessive lethal condition will lead to the death of individuals in a dihybrid F_2 in the proportion of :

(A) 1/4 (B) 1/2
(C) 3/4 (D) 7:6

93. The following amongst the molecular markers is co-dominant in nature :

(A) RAPD (B) VNTR
(C) SSR & RFLP (D) None of these

94. DNA can be denatured by heating at a temperature of :

(A) 72°C (B) 86°C
(C) 90°C (D) 94°C

95. From the marriage between two heterozygotes for albinism, what is the chance that the first would be non-albino?

(A) 1/4 (B) 1/2
(C) 3/4 (D) 3/8

96. Substitution haploids are represented as :

(A) n - 1 (B) n + 1
(C) n - 1 + 1 (D) None

97. The cDNA libraries can be prepared by isolating :

(A) rDNA (B) mRNA
(C) DNA (D) tRNA

98. The analogue of 5-bromouracil is :

(A) Adenine (B) Guanine
(C) Cytosine (D) Thyamine

99. Multiple strain or multi clone varieties are popular in :

(A) Forage grasses (B) Tomato
(C) Tobacco (D) Tapioca

100. The smallest unit of genetic material that can not be divided but is interchangeable among units refers to :

(A) Cistron (B) Muton
(C) Recon (D) Plasmon

101. Seedless watermelon is a :

(A) Diploid (B) Triploid
(C) Tetraploid (D) Pentaploid

102. In a cross if we get 8 types of gametes from both male and female parents, the resultant selfing will produce a ratio of :

(A) 1 : 1 : 1 : 1 : 1 : 1: 1 : 1
(B) 27 : 9 : 9 : 9 : 3 : 3 : 3 : 1
(C) 9 : 9 : 9 : 9 : 3 : 3 : 3 : 1
(D) 9 : 3 : 3 : 1

103. What is the plausible reason for availability of most of the tomato varieties in the market to be almost uniformly oval shaped and firm :

(A) Expansion in genetic diversity
(B) Cloning
(C) Reduction in genetic diversity
(D) Floral abnormality

104. Which of the event(s) given below is/are correct?

(A) The DNA synthesis take place in prophase stage of cell division
(B) The duration of prophase is larger than anaphase
(C) Both (A) and (B)
(D) Neither (A) nor (B)

105. Sugarcane is a :

(A) Autopolyploid
(B) Complex polyploid
(C) Auto allopolyploid
(D) Tetraploid

106. Which of the statements given below is/are correct?

(A) Dominance variance can be utilized for developing hybrids
(B) Additive genetic variance can be utilized to develop a variety
(C) Both (A) and (B)
(D) Neither (A) nor (B)

107. Considering the following landmark events in crop variety development in India :

1. Development of transgenic crops
2. Semidwarf varieties of wheat and rice
3. Hybrid cotton

Which one of the following is the correct chronological sequence of the above events?

(A) 1 - 3 - 2 (B) 2 - 3 - 1
(C) 2 - 1 - 3 (D) 3 - 2 - 1

108. Medium stable and extra long stable *Gossypium* lint are respectively produced from cultivars of :

(A) *G. hirsutum* and *G. barbadense*
(B) *G. herbaceum* and *G. barbadense*
(C) *G. arboreum* and *G. anomalum*
(D) *G. arboreum* and *G. hirsutum*

109. Pusa phalguni is a :

(A) Vegetable cowpea
(B) Brinjal variety
(C) Hybrid sunflower
(D) Seedless papaya

110. The immediate effect of pollen on the character of endosperm is called :

(A) Xenia (B) Metaxenia
(C) Hyperxenia (D) Epistasis

111. An individual with gametic chromosome number is :

(A) Haploid (B) Monoploid
(C) Dihaploid (D) Polyploid

112. The individuals with $2n\pm1$ chromosomes are commonly known as :

(A) Euploid (B) Aneuploid
(C) Heteroploids (D) None of these

113. Triticale is :

(A) Interspecific cross
(B) Intraspecific cross
(C) Intergeneric cross
(D) Intrageneric cross

114. An individual with 2n-1-1chromosome constitution is called :

(A) Monosomic (B) Disomic
(C) Double monosomic
(D) Double trisomic

115. The situation in which one gene hides the effect of a second gene when both are present in a chromosome is called :

(A) Dominance
(B) Epistasis
(C) Both (A) and (B)
(D) None of these

116. Flavr savr' is associated with :

(A) Potato (B) Pomato
(C) Tobacco (D) Tomato

117. Dr. C.R. Rao is a famous plant breeder working on :

(A) Sorghum (B) Maize
(C) Bajra (D) Ragi

118. *Arachis monticola* the most likely progenitor of groundnut is a :

(A) Diploid (B) Triploid

(C) Tetraploid (D) Hexaploid

119. Physical basis of life is :

(A) Nucleus (B) Protoplasm

(C) Cell (D) Food

120. Central Rice Research Institute at Cuttack was established during :

(A) 1946 (B) 1954

(C) 1972 (D) 1963

121. NBPGR is located at :

(A) New Delhi (B) Karnal

(C) Bikaner (D) Lucknow

122. The capacity of a cell to develop into a complete plant is called :

(A) Totipotency

(B) Fecundity

(C) Tissue culture

(D) Haploidy

123. 'Bulbosum Technique' is used in barley for the production of :

(A) Haploids (B) Monoploids

(C) Trisomics (D) Hybrids

124. Wheat is naturally :

(A) Auto polyploids

(B) Auto - allopolyploids

(C) Segmental allopolyploids

(D) Segmental autopolyploids

125. 'Colchicine' is generally used for :

(A) Hybridization

(B) Chromosome doubling

(C) Mutation

(D) All above

126. According to Vavilov, diploid wheats and hexaploid wheats have originated respectively from :

(A) Central Asia and Asia Minor

(B) Asia Minor and Central Asia

(C) Abyssinia and North America

(D) Both from Asia Minor

127. Sum total of deleterious genes present in species is known as :

(A) Genetic advance

(B) Genetic erosion

(C) Genetic load

(D) Species diversity

128. 'Chromosomal theory' of inheritance was given by :

(A) Sutton and Boveri

(B) Watson and Crick

(C) Beadle and Tatum

(D) Bateson and Punnet

129. 'Multiple factor hypothesis' is given by :

(A) East

(B) Sutton and Boveri

(C) Nilson Ehle

(D) Galton

130. Reverse transcription is also known as :

(A) Teminism (B) Temin

(C) Translation (D) All of these

131. Teosinte is the anscestor of maize was proposed by :

(A) Reeves (B) Beadle

(C) Manglesdorf (D) Gardner

132. Double helical structure of DNA was shown by :

(A) Jacob and Monod

(B) Beadle and Tatum

(C) Bateson and Punnet

(D) Watson and Crick

133. Translocation involves :

(A) Homologous chromosomes

(B) Non homologous chromosomes

(C) Chromosomes from other species

(D) None of these

134. The phenotype (P) is the function of genotype (G) and environment (E), but when individual genotype is subjected to multi-environmental testing, the phenotypic expression is the result of :

(A) G alone (B) E alone
(C) G + E (D) G + E + (G x E)

135. Mashuri is a *japonica* x *indica* variety developed in :

(A) Taiwan (B) Malaysia
(C) Indonesia (D) Singapore

136. When both the alleles of a gene are fully expressed in a heterozygote, what is this phenomenon called?

(A) Complete dominance
(B) Over dominance
(C) Co-dominance
(D) Pseudo-dominance

137. Centre of diversity refers to the area where cultivated plant species and/or their wild relatives show :

(A) High ecosystem diversity
(B) Highest variation
(C) Both (A) and (B)
(D) Low competition with unreleated species

138. All India Co-ordinated Rice Improvement Programme was established during :

(A) 1934 (B) 1948
(C) 1965 (D) 1954

139. Germplasm theory was proposed by :

(A) Weisman (B) Castle
(C) Darwin (D) Phillips

140. The phenomenon of linkage was first observed by :

(A) Hugo de Vries
(B) Morgan
(C) Bateson and Punnet
(D) Mendel

141. Sub species of rice suited to temperate region :

(A) *indica* (B) *japonica*
(C) *javanica*
(D) *javanica* and *japonica*

142. The paddy breeding station at Coimbatore was established in :

(A) 1921 (B) 1937
(C) 1952 (D) 1954

143. An amphidiploid between *B. nigra* and *B. campestris* is :

(A) *B. juncea* (B) *B. carinata*
(C) *B. napus* (D) *N. tomentosa*

144. During mitosis, chromosomal DNA duplication takes place in :

(A) Interphase (B) Prophase
(C) Metaphase (D) Anaphase

145. Lysine is limiting amino acid in :

(A) Cereals (B) Pulses
(C) Oil seeds (D) None of these

146. Protein involved in keeping the DNA single stranded during replication is :

(A) Topoisomerase
(B) Helicase II
(C) Ligase
(D) SSB protein

147. Who made first intergeneric cross between bread wheat and rye?

(A) Richey (1923)
(B) Muntzing (1951)
(C) Vavilov (1926)
(D) Rimpu (1928)

148. The diploid number of corn is 20. How many chromosomes are expected in the product of the second karyokinesis following meiosis in the formation of an embryo sac?

(A) 10 (B) 20
(C) 30 (D) 40

149. Compared to other races photoperiodic response in *javanica* rice is :

(A) Very sensitive
(B) Less sensitive
(C) More sensitive
(D) As sensitive as other races

150. ------------- in Ladino clover is governed by two dominant alleles, one produces cyanogenic glucoside, while the other releases the enzyme to catalyze its release from glucoside.

(A) Oxalic acid (B) IBA
(C) NAA (D) HCN

151. The name of the agriculture scientist, who was awarded the Nobel peace prize is :

(A) Dr. K.C. Naik
(B) Sir C.V. Raman
(C) Dr. M.S. Swaminathan
(D) Dr. Norman E. Borlaug

152. Self-incompatibility is controlled by :

(A) Duplicate factors
(B) Polygenes
(C) Multiple alleles
(D) Plasmogenes

153. By gene regulations, it is generally understood that genes regulate phenotypic expression through :

(A) Mutation (B) Cell division
(C) Protein (D) Macro molecules

154. "If a gene is represented in an infinitely large random mating population by adaptively neutral alleles, A and a, in the ratio of qA : (1-q)a, the genotypic proportions : $q^2AA + 2q(1-q)Aa + (1-q)^2 aa = 1$" is known as :

(A) Hardy-Weinberg law
(B) Weibe-Thomas law
(C) Briggs-Harlan law
(D) Stanford-Suneson law

155. The study of segregation of character pair led Gregor Mendel to postulate law of inheritance in :

(A) 1900 (B) 1865
(C) 1906 (D) 1903

156. 2n - 1 is a :

(A) Alloploid (B) Trisomic
(C) Monosomic (D) Tetrasomic

157. Which of the following cells is normally diploid?

(A) Primary polar body
(B) Primary spermatocyte
(C) Spermatid
(D) Spermatozoa

158. Ratio of the genetic variance to the total variance which determines the relationship between phenotype and genotype is known as :

(A) Heritability (B) Genetic advance
(C) Penetrance (D) Dominance

159. Mendel did not face the problem of linkage, however, seven characters studied by him were located on how many chromosomes?

(A) Three (B) Four
(C) Five (D) Seven

160. Test cross ratio in case of duplicate gene action will be :

(A) 1 : 1 (B) 1 : 1: 1: 1
(C) 1 : 3 (D) 3 : 1

161. The chromosome are most condensed at :

(A) Prophase (B) Metaphase
(C) Anaphase (D) Telophase

162. Cell linkage theory was proposed by :

(A) Hooke
(B) Nagali
(C) Schlieden and Schwan
(D) Virchow

163. Sex in plants was discovered by :
(A) T Fairchild (B) Camerarious
(C) Grew (D) Linnaeus

164. Who gave the concept of circular chromosome in *E. coli*?
(A) Cairns (B) Meselson
(C) Watson (D) Stahl

165. Number of chromosomes in wheat endosperm is :
(A) 28 (B) 63
(C) 42 (D) 14

166. The term gene was coined by :
(A) Mendel (B) Johannsen
(C) Shull (D) Bateson

167. A : T ratio in one strand of DNA duplex is 2, what will be the ratio in the complementary strand?
(A) 0.5 (B) 2.0
(C) 2.5 (D) 1.0

168. Site of protein synthesis in a cell is :
(A) Ribosomes
(B) Mitochondria
(C) Chloroplasts
(D) Endoplasmic reticulum

169. Which cell organelle is having digestive enzymes?
(A) Ribosomes (B) Lysosomes
(C) Golgi bodies (D) Mitochondria

170. Transgressive variation results in :
(A) Accumulation of dominant genes
(B) Interaction of polygenes
(C) Introgression of major genes
(D) Increase in the mean of the population

171. The defective copies of normal genes are known as :
(A) Pseudo genes (B) Jumping genes
(C) Split genes (D) Selfish genes

172. The terms recon, muton and cistron were coined in 1957 by :
(A) Benzer (B) Muller
(C) Bridges (D) Morgan

173. If individuals of genotype $G^1G^2H^1H^2J^1J^2$ are intercrossed, how many phenotypes can occur in their offspring?
(A) 6 (B) 12
(C) 16 (D) 27

174. The variance which is fixable is :
(A) Additive (B) Dominance
(C) Epistatic (D) All above

175. If the genetic composition of a genotype is AbC, Abc, abc, abC in equal proportion, the genotype of the source individual is :
(A) AaBbCc (B) AabbCc
(C) AABbCc (D) AabbCC

176. IPGRI headquarter is situated in :
(A) Rome, Italy
(B) New York, USA
(C) New Delhi, India
(D) Sydney, Australia

177. Golgi apparatus is also known as :
(A) Ribosomes (B) Dictyosome
(C) Centriole (D) Lysosome

178. 'Law of homologus series' is given by :
(A) Biffen (B) Vavilov
(C) Debary (D) Darwin

179. In the DNA double helix G/A is equal to :
(A) T/A (B) C/T
(C) C/A (D) C/G

180. Colchicine' interferes with :
(A) Replication
(B) Recombination
(C) Formation of spindle fibres
(D) Amplification

181. When chiasma frequency is 100%, the percentage of crossing over is :

(A) 25% (B) 50%
(C) 100% (D) 150%

182. The minimum progeny population size allowing for random union of all kinds of gametes from AaBbCc parent is :

(A) 9 (B) 27
(C) 64 (D) >100

183. An organelle present in animal cell but missing from plant cell is :

(A) Nucleolus (B) Centriole
(C) Vacuole (D) Mitochondria

184. The 'one gene one enzyme' hypothesis was proposed by :

(A) Beadle and Tatum
(B) Lederberg and Tatum
(C) Watson and Crick
(D) Muller and Stadler

185. If parents differ in ten gene pairs, no selection is done; six backcrosses will produce a population having ------------- per cent homozygous and identical individuals to recurrent parent at all ten loci.

(A) 65 (B) 75
(C) 85 (D) 95

186. The common form of DNA present in the living organisms is the :

(A) A form (B) B form
(C) C form (D) Z form

187. Source of dwarfing gene in wheat is :

(A) Opaque (B) Norin - 10
(C) Dee - Gee - Woo - Gen
(D) Hipsoli

188. If the genetic endowments of two nuclei that unite to produce the plant zygote are labelled A and B, and the other product of fertilization within that same embryo sac is labelled ABB, then the tube nucleus that was in the pollen tube that delivered the fertilizing male gamates must be labeled as:

(A) A (B) B
(C) AB (D) BB

189. The concept of physiological races was given by :

(A) De Barry (1914)
(B) Barrus (1911)
(C) Robinson (1971)
(D) Russell (1963)

190. The source of 'A' genome in wheat is :

(A) *T. urartu* (B) *T. dicoccoides*
(C) *T. durum* (D) *T. turgidum*

191. DNA polymerase I was discovered by :

(A) Harshey (B) Yanofsky
(C) Crick (D) Korenberg

192. BTT is associated with :

(A) Wheat (B) Rice
(C) Barley (D) Maize

193. Triploids are normally :

(A) Sterile (B) Male sterile
(C) Fertile (D) None of these

194. First man made cereal is :

(A) Triticale (B) Raphanobrasica
(C) Pomato (D) None of these

195. Cells without cell wall are known as :

(A) Proplastids (B) Protoplasts
(C) Chloroplasts (D) Cybrids

196. The number of bases in the single turn of the DNA helix is :

(A) 5 (B) 6
(C) 10 (D) 17

197. Inhibitors to translation (Protein synthesis) are :

(A) Chloramphanicol and Erythromycin
(B) Gibberelin
(C) Ethylene
(D) ABA

198. The dwarfing gene in rice is

(A) Dee Gee Woo Gen
(C) Norin 10
(C) Milo
(D) Tift

199. The primary centre of origin of Hordeum Vulgare (Barley) is

(A) Abyssynian centre
(B) Asia Minor centre
(C) Mediterranean centre
(D) Central Asia centre

200. --------------- in Ladino clover is governed by two dominant alleles, one produces cyanogenic glycoside, while the other releases the enzyme to catalyze its release from gloycoside

(A) Oxalic acid (B) IAA
(C) HCN (D) NAA

201. 2n-1 is a

(A) Monosomic (B) Transomic
(C) Tetrasomic (D) Alloploid

202. A cell or an organism having one chromosome pair in addition to the normal somatic compliment of the species is

(A) Monosomic (B) Nullisomic
(C) Tetrasomic (D) Trisomic

203. A character determined by many genes and does not show discrete variation is known as

(A) Qualitative character
(B) Oligogenic character
(C) Quantitative character
(D) Multiple-allelic character

204. A criss-cross inheritance is shown by

(A) Sex-influenced character
(B) Sex-limited character
(C) Sex-linked character
(D) Sex-quantitative character

205. A mendelian population means

(A) Self-pollinated population
(B) Inbred population
(C) Random-mating population
(D) Hybrid population

206. A plant with 2n=20 will have how many linkage group

(A) 20 (B) 10
(C) 40 (D) 5

207. A plant with genotype Gg Ww will produce gametes of which of the following types?

(A) GW, gw, gW, Gw
(B) GG, WW, gg, ww
(C) Gw, GG, GW
(D) gg ww

208. Albinism in plants is associated with

(A) Epistasis (B) Recessive lethal
(C) Dominant lethal
(D) Chromosome duplication

209. Alternative forms of a gene is known as

(A) Mutation (B) Gamete
(C) Allele (D) Genotype

210. An endospermic cell of a seed after somatic chromosome number (2n=20) will have

(A) 30 Chromosomes
(B) 20 Chromosomes
(C) 10 Chromosomes
(D) 40 Chromosomes

211. An individual having different alleles for any gene pair and producing two kinds of gametes is known as

(A) Hemizygous (B) Homozygous
(C) Heterozygous (D) Heterogenous

212. An organism having the gametic chromosome numbers is called as

(A) Genome (B) Hybrid
(C) Gamete (D) Haploid

213. Bread wheat is

(A) Haploid (B) Diploid
(C) Tetraploid (D) Allohexaploid

214. Carolus Linnaeus is credited with introducing the

(A) Law of limiting factors
(B) Lawof independent assortment
(C) Binomial system of nomenclature
(D) Laws of inheritance

215. Who is regarded as the father of botany?

(A) Theofrastus
(B) Gregor Mendel
(C) Carolus Linnaeus
(D) Oswald Tippo

216. Chromosome numbers in somatic cells of a maize plant are

(A) 22 (B) 26
(C) 14 (D) 20

217. Chromosome which pair at meiotic prophase and are similar size, shape, structure and functions are known as

(A) Homologous
(B) Isocentric
(C) Isochromosome
(D) Acentric

218. Clone in general is

(A) Homozygous (B) Heterogeneous
(C) Heterozygous (D) Often heterozygous

219. Crossing over occurs at

(A) Two strand stage
(B) Four strand stage
(C) Three strand stage
(D) None of the above

220. Cytogenetically, the definitive nucleus is

(A) Haploid (B) Diploid
(C) Triploid (D) Tetraploid

221. Who invented first compound microscope?

(A) Robert Hooke (B) Jansen
(C) Leeuwenhoek (D) None of these

222. Electron microscope was invented by

(A) Jansen
(B) Leeuwenhoek
(C) Knoll and Ruska
(D) None of the above

223. Epitasis should be carefully distinguished from dominance, which refers to non-additivity of alleles at

(A) The same locus of a homologous chromosome
(B) Different loci of a homologous chromosome
(C) The same locus of non-homologous chromosomes
(D) Different locus of non-homologous chromosomes

224. Exchange of genetic material takes place between non-sister chromatids of homologous chromosomes during the

(A) Four stand stage in prophase I
(B)Two strand stage in prophase I
(C) Metaphase I stage in meiosis
(D) Two strand stage in prophase

225. Who established that character is transmitted from generation to generation through chromosomes?

(A) Robert Brown
(B) Robert Hooke
(C) Nawaschin
(D) Strasburger and Hertwig

226. Genes are made up of

(A) RNA only (B) DNA only
(C) RNA & DNA (D) Proteins

227. Genetics is the study of

(A) Genes
(B) Gene interaction
(C) Heredity and variation
(D) DNA

228. Genotype and Environment interaction would take place when there is / are

(A) One genotype and two environments
(B) Two genotype sand one environment
(C) Three genotypes and one environment
(D) Two or more genotypes and two or more environments

229. Given that the somatic chromosome number of *Triticum aestivum* is 2n = 6x = 42, which one of the following pairs is correctly matched?

(A) Monosome (B) Trusine
(C) Tetrasome (D) Nullisome

230. Mendel is famous for

(A) Propounding the laws of inheritance
(B) Laws of limiting factors
(C) The theory of mutation
(D) The discovery of penicillin

231. Group of plants with similar characteristics of any rank is termed

(A) Taxon (B) Species
(C) Genus (D) Order

232. Heritability may be defined as the

(A) Interaction product of genotype with present in a species
(B) Sumtotalof hereditary material environment
(C) Degree of resemblance between the original and the selected plants
(D) Proporation of phenotypic variability, which is due to heredity

233. In sweet pea gene C or P alone produces white flower, whereas both C and P are needed to produce purple coloured flower. In a particular cross between two white flowered plants the offsprings were one half purple and one-half white. The possible genotypes of the parents is

(A) Cc Pp x ccPp (B) CCpp x ccPP
(C) ccPp x CCpp (D) Ccpp x ccPp

234. In wheat and oat,there are ----------- genes for red versus white kernels

(A) Two (B) Three
(C) Four (D) Five

235. Magnification of an object by electron microscope is

(A) 1,000 X (B) 10,000 X
(C) 1,00,000 X
(D) More than 2,00,000 X

236. Monoploid is also a

(A) Haploid (B) Diploid
(C) Monosomic (D) Triploid

237. Pink (NN) flowered plants crossed with blue (nn) flowered plants gave black (Nn) flowered plants in F1 generation. If F1 plants are selfed, which of the following will be the phenotype of F2 progeny?

(A) All blue
(B) ½ blue and ½ pink
(C) All pink
(D) ¼ pink, ½ black, ¼ blue

238. Which of the following scientists is called the father of genetics?

(A) Charles Darwin
(B) Robert Hooke
(C) Gregor Mendel
(D) Gregor Morgon

239. Polygenes affecting the same trait, with each enhancing the phenotype are termed as

(A) Amnions
(B) Alcaptonuria
(C) Allosteric effect
(D) Additive factors

240. Population improvement is used

(A) To accumulate deleterious alleles
(B) To maintain homozygosity
(C) To accumulate favourable alleles
(D) To increase heterozygosity

241. Random assortment of genes will occur if the gene are located on

(A) Same chromosome

(B) Different chromosomes

(C) Sex chromosome

(D) Autosome

242. Recessive lethal genes are carried in the

(A) Heterogenous form

(B) Hetrozygous form

(C) Homozygous form

(D) Homogenous form

243. Sex in Drosophila is governed by the following mechanism

(A) XY methods (B) OX method

(C) ZZ method (D) X autosomes ratio

244. Superiority of heterozygote leads to

(A) Fixation of recessive allele

(B) Fixation of dominant allele

(C) Maintenance of both the alleles

(D) Depression of inbreeding

245. Tetraploid potato has 48 chromosomes, whereas ordinary piploid species have ----------- chromosomes

(A) 34 (B) 24

(C) 14 (D) 10

246. The basis on which evolution works is

(A) Simplicity (B) Complexity

(C) Variations (D) Adaptation

247. The best-known artificial system of classification if that promulgated by

(A) Sir Julian Huxley

(B) Bentham and Hooker

(C) Linnaeus (D) A.P.Candolle

248. The best method to break the unfavourable linkage is

(A) Single seed descent method

(B) Diallel selective mating

(C) Pedigree method

(D) Bulk method

249. The branch of genetics concerned with the study of frequencies of mating populations is

(A) Quantitative genetics

(B) Behavioural genetics

(C) Population genetics

(D) Eugenetics

250. The evolutionary history of a species is referred as

(A) Phylogeny (B) Progeny

(C) Organic evolution

(D) Natural selection

251. The genes that intensify or diminish the effect of major genes are known as factors.

(A) Modifying (B) Nullifying

(C) Falsifying (D) None of these

252. The main reason for the classification of plants is

(A) The difficulties arising in the study of plants

(B) Their mechanical separation because of their large number

(C) The nature of man

(D) To identify the plants

253. The mendel's laws of inheritance were rediscovered in 1900 by

(A) Bateson and Punnet

(B) Beadle, Tatum and Leader- berg

(C) Watson and Crick

(D) Correns, de Varies and Tschermak

254. The normal eye colour in Drosphila is red. But mutants occur having white eye and also different shades ranging between white and red which are all recessive to red, while white colour is recessive to all others. This phenomenon is due to

(A) Pseudodominance

(B) Pseudoallelism

(C) Codominance

(D) Polygenes

255. The nucleus of chromosome in common wheat is

(A) 36 (B) 38
(C) 48 (D) 42

256. The physical basis of heredity is termed as

(A) Gene pool (B) Gene frequency
(C) Germplasm (D) Genetic gain

257. The process of making the expression of a non-allele gene by another gene or gene pair is known as

(A) Epistasis
(B) Hypostasis
(C) Nobilisation
(D) None of the above

258. The proportion of one allele as represented in a breeding population is known as

(A) Gene frequency (B) Gene pool
(C) Genetic gain (D) Genetic resource

259. The science dealing with variation and heredity is called

(A) Immunology (B) Cytology
(C) Genetics (D) Radiology

260. The science, which deals with the study of interrelationship between the living organisms and the various factors of the environmental surrounding them as

(A) Botany (B) Biology
(C) Ecology (D) Zoology

261. The study of ecology from the standpoint of individual species is called

(A) Synecology (B) Phytogeography
(C) Autecology (D) Allogamy

262. The term epistasis was originally used by ---------------- in 1909 to describe genes whose effects mark or cover the effects of other genes

(A) Wagner (B) Bateson
(C) Hayman (D) H.J. Mujer

263. The term genetics was coined by

(A) Gregor Mendel
(B) Bateson
(C) T.H. Morgan
(D) H.J. Mujer

264. The term used in sugarcane breeding to denote the crossing of *Sacharum officinarum* with related species followed by one or more back crosses is called

(A) Mass selection
(B) Crossing over
(C) Cane improvement
(D) Nobilisation

265. The theory explaining the cause of variation and evolution, (Continuity of Germplasm) was put forward by

(A) Charles Darwin
(B) Gregor John Mendel
(C) Weismann
(D) Hugo de Vries

266. The theory of catastrophism was introduced by

(A) Christian Church
(B) Cuvier
(C) Lmark
(D) Darwin

267. The theory of genes in the chromosomes was introduced by

(A) Mendel (B) Morgan
(C) Strasburger (D) Hugo de Vries

268. Theory of 'Acquired characters' was proposed by

(A) Darwin (B) Lamark
(C) Bateson (D) Weismann

269. Theory of 'Pangenesis' was given by

(A) Hugo de Vries
(B) Charles Darwin
(C) Lamark
(D) Bateson

270. There are four stage of development of Genetics upto 20th Century. The correct order of their occurrence is

(A) Darwinism, Lamarckism, Mendelism, Weismannism

(B) Lamarckism, Darwinism, Weismannism, Mendelism

(C) Lamarckism, Weismannism, Darwinism, Mendelism

(D) Weismannism, Lamarckism, Darwinism, Mendelism

271. When two genes show independent segregation, a genotype AaBb will produce ---------- types of gametes :

(A) 4 (B) 3

(C) 2 (D) 1

272. When two alleles of a gene are identical, the plant is said to be

(A) Recessive (B) Dominant

(C) Heterozygous (D) Homozygous

273. Transmission of characteristics and qualities of parents of their offspring is known as

(A) Mutation (B) Heredity

(C) Adaptation (D) Hybridization

274. When a formation is dominated by one or more species which are abundant in it, is said to be

(A) Association (B) Climax

(C) Community (D) population

275. When a plant heterozygous for two given characters Tt Rr, both of them are completely dominant, is self-pollinated, the resulting progeny will consist of

(A) Six phenotypes

(B) Four phenotypes

(C) Three phenotypes

(D) Two phenotypes

276. When a Tall plant (Tt) in pea is crossed with another Tall plant (Tt), the probability of obtaining a dwarf plants is

(A) 1:2 (B) 1:4

(C) 1:8 (D) 1:16

277. Multiple effects of a single gene is known as

(A) Polyploidy (B) Heteroploidy

(C) Pleiotropy (D) None of these

278. The test cross ratio in complementary gene action is

(A) 3:1 (B) 1:3

(C) 1:2:1 (D) 2:1:1

279. Triticale is

(A) Interspecific cross

(B) Intraspecific cross

(C) Intergeneric cross

(D) intrageneric cross

280. Bread wheat is

(A) Monoploid (B) Diploid

(C) Tetraploid (D) Hexaploid

281. Ratio of additive variance to phenotypic variance is called :

(A) Heritability (Broad sense)

(B) Heritability (Narrow sense)

(C) Co-heritability

(D) Gene Action

282. A Nullisomic individual is represented by :

(A) 2n-1 (B) 2n+1

(C) 2n-2 (D) 2n+2

283. After the cell has undergone mitosis, the amount of DNA in daughter cells will be

(A) Equal to the parent cell

(B) Half of the parent cell

(C) Double of the parent cell

(D) Triple of the parent cell

284. In cereals, limiting amino acids are

(A) Lysine

(B) Arginine

(C) Methionine

(D) Thayamin

285. In green vegetables, the limiting amino acids is:

(A) Lysine (B) Arginine
(C) Methionine (D) Tryptophan

286. High malting barley includes:

(A) Pusa 12 (B) Kufri 26
(C) Karan 280 (D) PAU 15

287. A Rice plant with 2n=24 will have how many linkage group

(A) 24 (B) 12
(C) 6 (D) 48

288. Mohan variety of rice is tolerant to:

(A) Drought (B) Salinity
(C) Metal toxicity (D) Acid soil

289. Akashi variety of rice is tolerant to:

(A) Metal toxicity
(B) Salinity
(C) Drought
(D) Acid soil

290. ICRISAT refers to

(A) International Crop Research Institute for Semi Arid Tropics
(B) Indian Crop Research Institute for Semi Arid Tropics
(C) Institutional Central Research Initiative for Semi Arid Tropics
(D) Indian Central Research Institute for Semi Arid Tropics

ANSWER SHEET

1. B	2. C	3. D	4. B	5. C	6. C	7. A	8. D	9. D	10. B
11. B	12. B	13. C	14. C	15. C	16. A	17. A	18. C	19. B	20. C
21. D	22. D	23. B	24. A	25. B	26. A	27. D	28. B	29. D	30. D
31. D	32. B	33. D	34. C	35. C	36. C	37. B	38. D	39. C	40. B
41. C	42. A	43. D	44. D	45. A	46. A	47. B	48. C	49. A	50. B
51. A	52. B	53. B	54. D	55. D	56. B	57. A	58. A	59. C	60. C
61. D	62. A	63. C	64. B	65. C	66. D	67. C	68. A	69. C	70. C
71. B	72. A	73. C	74. B	75. D	76. A	77. A	78. A	79. D	80. B
81. D	82. D	83. C	84. A	85. D	86. B	87. B	88. C	89. A	90. D
91. A	92. A	93. C	94. D	95. C	96. C	97. B	98. D	99. A	100. C
101. B	102. B	103. D	104. B	105. B	106. C	107. B	108. A	109. A	110. A
111. A	112. B	113. C	114. C	115. B	116. D	117. A	118. C	119. A	120. A
121. A	122. A	123. A	124. C	125. B	126. B	127. C	128. A	129. C	130. A
131. C	132. D	133. B	134. D	135. B	136. C	137. C	138. C	139. A	140. C
141. B	142. A	143. A	144. A	145. A	146. D	147. D	148. D	149. B	150. D
151. D	152. C	153. C	154. A	155. B	156. C	157. B	158. A	159. B	160. D
161. C	162. D	163. B	164. A	165. B	166. B	167. A	168. A	169. B	170. D
171. A	172. A	173. D	174. A	175. B	176. A	177. B	178. B	179. B	180. C
181. B	182. C	183. B	184. A	185. C	186. B	187. B	188. A	189. B	190. A
191. D	192. C	193. A	194. A	195. B	196. C	197. A	198. A	199. A	200. C

201. **A**	202. **A**	203. **D**	204. **C**	205. **C**	206. **B**	207. **A**	208. **A**	209. **C**	210. **C**
211. **B**	212. **D**	213. **D**	214. **C**	215. **A**	216. **D**	217. **A**	218. **C**	219. **B**	220. **B**
221. **B**	222. **C**	223. **B**	224. **A**	225. **D**	226. **B**	227. **C**	228. **B**	229. **C**	230. **A**
231. **A**	232. **D**	233. **C**	234. **B**	235. **D**	236. **A**	237. **D**	238. **C**	239. **D**	240. **C**
241. **B**	242. **B**	243. **B**	244. **A**	245. **B**	246. **C**	247. **C**	248. **B**	249. **C**	250. **A**
251. **A**	252. **A**	253. **D**	254. **A**	255. **D**	256. **A**	257. **A**	258. **A**	259. **C**	260. **C**
261. **C**	262. **C**	263. **B**	264. **D**	265. **B**	266. **B**	267. **B**	268. **B**	269. **B**	270. **C**
271. **A**	272. **D**	273. **B**	274. **B**	275. **B**	276. **B**	277. **C**	278. **B**	279. **C**	280. **D**
281. **B**	282. **C**	283. **A**	284. **A**	285. **C**	286. **C**	287. **B**	288. **B**	289. **C**	290. **A**

Chapter 2

PLANT DOMESTICATION

1. What is domestication?
   (A) Bringing wild species under animal management
   (B) Bringing wild species under wildlife management
   (C) Bringing wild species under human management
   (D) Bringing wild species under stress management

2. For domestication, the wild species will have changed ---------
   (A) Totally (B) Partially
   (C) Fifty percent
   (D) Twenty five percent

3. In which crop, the asexual reproduction has been promoted under plant domestication?
   (A) Cotton (B) Groundnut
   (C) Neem (D) Sugarcane

4. Which one of the following is preferred for domestication?
   (A) Aneuploidy (B) Polyploidy
   (C) Autogamy (D) Polygamy

5. Give an example for shortening of lifecycle under domestication/
   (A) Redgram (B) Sugarcane
   (C) Jute (D) Wheat

6. In which of the following crops, the plant height has increased due to domestication
   (A) Redgram (B) Sugarcane
   (C) Jute (D) wheat

7. In -----------, the plant height was decreased under domestication
   (A) Pulses (B) Oilseeds
   (C) Fibre crops (D) Cereals

8. In which family, the bitter principle of species has been decreased due to domestication?
   (A) Malvaceae (B) Cucurbitaceae
   (C) Compositae (D) Amaranthaceae

9. The principle of dormancy has been completely eliminated in the breeding method of -------------
   (A) Plant introduction
   (B) Plant hybridization
   (C) Plant domestication
   (D) Plant selection

10. Which one of the following is incorrect with respect to plant domestication
    (A) Shattering of pods reduced
    (B) Flowering is eliminated
    (C) Tillering altered
    (D) Plant height reduced

ANSWER SHEET

1. C	2. A	3. D	4. B	5. A	6. C	7. D	8. A	9. C	10. B

Chapter 3

PLANT INTRODUCTION

1. Acclimatisation is :
   (A) A process of adjustment by crop plants to environment
   (B) Removal of female parts of a flower
   (C) Related with the clay
   (D) Changing of climatical condition
2. Characteristic feature of the hydrophyte is :
   (A) Spongy stem
   (B) Feebly developed root-system
   (C) Reduced protective tissues
   (D) All the above
3. ADT 27 is :
   (A) Inter racial variety
   (B) Inter specific variety
   (C) Inter generic variety
   (D) A mutant variety
4. Fibre length in cotton is measured by the instrument :
   (A) Areolometer (B) Micronaire
   (C) Balls sorter (D) Pressley tester
5. China rose is a :
   (A) Herb (B) Tree
   (C) Shrub (D) Climber
6. CIMMYT is located at :
   (A) Mexico (B) Nigeria
   (C) India (D) Spain
7. Conifers are abundant in :
   (A) Tropical zone (B) Temperate zone
   (C) Alpine zone (D) Subtropical zone
8. Cotton :
   (A) Is a cross pollinated crop
   (B) Is a self pollinated crop
   (C) Is an often-cross pollinated crop
   (D) Is an often-self pollinated crop
9. Quality cotton should have :
   (A) Long staple, low micronaire and strong fibre
   (B) Long staple, strong fibre and high micronaire
   (C) Medium staple, strong fibre and low micronaire
   (D) Medium staple, low micronaire and high spinning value
10. The inherent capacity of an organism to increase in numbers under ideal condition is known as :
    (A) Growth rate
    (B) Reproductive potential
    (C) Biotic potential
    (D) Population explosion
11. Two taxonomic species are distinguished from each other by :
    (A) Discontinuity in a set of correlated characters
    (B) Their ability to exchange genes freely
    (C) Their failure to interbreed
    (D) Their similarity in correlated morphological characters

12. The ratio of economic yield to the biological yield is :

(A) Leaf area ratio (B) Total harvest
(C) Harvest index (D) Total biomass

13. Shaktiman is the new variety of :

(A) Chickpea (B) Rice
(C) Wheat (D) Maize

14. The low neurotoxin variety of *Lathyrus* is :

(A) Pusa 12 (B) Pusa 9
(C) Pusa 24 (D) Pusa 26

15. Land races are :

(A) Local varieties
(B) Hybrid varieties
(C) Advanced strains
(D) Wild Species

16. Electron microscope was invented by :

(A) Knoll and Ruska
(B) Leeuwenhoek
(C) Jansen
(D) Johnson

17. In China rose flower is :

(A) Hypogynous (B) Perigynous
(C) Epigynous (D) Clestiogamy

18. In asparagus one or more short, green cylindrical or sometimes flattened branches of limited growth develop from node of stem, such a branch is known as :

(A) Cladode (B) Thorn
(C) Stem tendril (D) Phylloclade

19. Rice varieties IR 20, IR 26 and IR 36 have shown better resistance to :

(A) Tungro (B) Brown plant hopper
(C) Blast (D) Bacterial leaf blight

20. Hydrilla is a :

(A) Hydrophyte (B) Hygrophyte
(C) Mesophyte (D) Xerophyte

21. Aruna is a :

(A) Popular mutant castor variety
(B) A hybrid castor
(C) Mutant sesamum variety
(D) Improved castor variety

22. Which of the following is incorrectly related?

(A) CIAT - Lima, Peru
(B) ILCA - Addis Ababa, Ethiopia
(C) WARDA - Monrovai, Liberia
(D) ILRAD - Nairobi, Kenya

23. Any changes in genetic composition of a population such as a change in gene frequency :

(A) Evolution
(B) Acclimatization
(C) Natural selection
(D) Genetic equilibrium

24. Adjustment of individual to their environment by mean of special structures or of function is spoken of :

(A) Mutation (B) Adaptation
(C) Heredity (D) Variation

25. Halophytes growing on muddy swamps form a special vegetation known as :

(A) Evergreen forests
(B) Riparian forests
(C) Mangrove forests
(D) Coniferous forests

26. A large unit of natural vegetation in an area under identical condition (climatic) is called :

(A) Climax
(B) Association
(C) Plant formation
(D) Community

27. Ponni paddy is :

(A) Inter spacial variety
(B) Inter generic variety
(C) Inter racial variety
(D) Intra racial variety

28. Consider the following genotypes
(1) Breeding lines, (2) Land races,
(3) Variety under cultivation
What is the correct sequence of these in terms of their importance to farmers
(A) 1, 2 and 3 (B) 3, 1 and 2
(C) 2, 3 and 1 (D) 1, 3 and 2

29. Fibres of economic importance from jute plant includes :
(A) Phloem only
(B) Xylem only
(C) Phloem and cortex
(D) Xylem and Phloem

30. The International Rice Research Institute was established at Philippines in :
(A) 1902 (B) 1962
(C) 1965 (D) 1973

31. Semi dwarf varieties of wheat are :
(A) More responsive to fertilizer application than tall varieties
(B) Less responsive to fertilizer application than the tall varieties
(C) Lodging easily than tall varieties
(D) Less tolerance to pest and disease than tall varieties

32. Origin of species' was published in :
(A) 1759 (B) 1859
(C) 1895 (D) 1959

33. The plant type concept was first introduced by :
(A) Jenings (1964)
(B) Donald (1968)
(C) Finley (1968)
(D) Rasmusson (1987)

34. A complex of communities of plants and animals of a region existing under identical climatic conditions is called as
(A) Biome
(B) Ecosystem
(C) Ecological complex
(D) Environment

35. Binomial system of Nomenclature means that the name of plant is made up of two words, which designate
(A) Family and genus
(B) Order and family
(C) Species and family
(D) Genus and species

36. Conifers are abundant in
(A) Tropical zone
(B) Alpine zone
(C) Temperate zone
(D) Subtropcial zone

37. Centre of diversity refers to the area where cultivated plant species and or their wild relatives show
(A) Low competition with unrelated species
(B) Highest variation
(C) High ecosystem diversity
(D) Highest variation and high ecosystem diversity

38. A large unit of natural vegetation in an area under identical condition (cimatic) is called
(A) Plant formation
(B) Association
(C) Community
(D) Climax

39. Acclimatisation is
(A) A process of adjustment by crop plants to environment
(B) Related with the clay
(C) Removal of female parts of a flower
(D) Changing of climatical condition

40. Adjustment of individual to their environment by mean of special structures or of function of spoken of
(A) Mutation (B) Variation
(C) Heredity (D) Adaptation4

41. Introduction of plant from western countries is called as ----------

(A) Indigenous introduction
(B) Foreign introduction
(C) Indigenous collection
(D) Country to country collection

42. Plant introduction from one state to another state is named as ---------

(A) Indigenous introduction
(B) Foreign introduction
(C) Foreign collection
(D) State to district introduction

43. Shifting of plants from native environment to new environment is called as --------

(A) Plant domestication
(B) Plant collection
(C) Plant introduction
(D) Germplasm collection

44. Which one of the following crop is introduced from other country?

(A) Maize (B) Cotton
(C) Ragi (D) Wheat

45. Which variety of wheat was introduced to India?

(A) IR 8 (B) IR 20
(C) Sonara 64 (D) TN 1

46. Which one of the following variety of rice was introduced to India?

(A) IR8 (B) Sonara 64
(C) Lerma Rojo (D) Jawahar

47. Which one of the following wheat variety was introduced from CIMMYT

(A) Kalyan sona (B) Sonara 64
(C) Jawahar (D) Lerma Rojo

48. Which organization is involved in the introduction of a variety from exotic source?

(A) IARI (B) ICRISAT
(C) NBPGR (D) ICAR

49. Where NBPGR is located?

(A) Calcutta (B) Pune
(C) Coimbatore (D) New Delhi

50. Which one of the following prefix is denoted for exotic collection?

(A) IC (B) IW
(C) EC (D) IB

51. To which category of planting material the quarantine is done for plant introduction?

(A) State to state
(B) Country to country
(C) State to district
(D) District to district

ANSWER SHEET

1. **A**	2. **D**	3. **A**	4. **C**	5. **C**	6. **A**	7. **B**	8. **C**	9. **B**	10. **B**
11. **C**	12. **C**	13. **D**	14. **C**	15. **A**	16. **A**	17. **A**	18. **A**	19. **B**	20. **A**
21. **A**	22. **A**	23. **A**	24. **B**	25. **C**	26. **C**	27. **C**	28. **C**	29. **D**	30. **B**
31. **A**	32. **B**	33. **A**	34. **A**	35. **D**	36. **C**	37. **D**	38. **A**	39. **A**	40. **D**
41. **B**	42. **A**	43. **C**	44. **D**	45. **C**	46. **A**	47. **A**	48. **C**	49. **D**	50. **C**
51. **B**									

Chapter 4

BREEDING METHODS FOR SELF POLLINATED CROPS

4.1. SELECTION

1. Pure line variety is a :
   (A) Homozygous - Heterogenous
   (B) Homozygous - Homogenous
   (C) Heterogenous - Heterogenous
   (D) Heterogenous - Homozygous
2. The crop that provides best documented example of long time effects of selection :
   (A) Sweet sorghum
   (B) Sugar beet
   (C) Sugar maple
   (D) Sugarcane
3. Earliest method of plant breeding :
   (A) Introduction (B) Selection
   (C) Hybridization (D) Mutation
4. Improvement of coconut is difficult due to:
   (A) Practical possibility of maintaining pure lines because of the factors of time and space
   (B) Increase in vigour in the selfed progenies
   (C) Extensive self pollination
   (D) Reduction in vigour of selfed progenies
5. ------------- provides the best documented example of long time effects of selection.
   (A) Sugarcane (B) Sugar maple
   (C) Sugarbeet (D) Sweet sorghum
6. Origin of species by means of Natural Selection's was given by
   (A) Lamark (B) Linnaeus
   (C) Charles Darwin (D) Erasmus
7. Recurrent selection is mostly used in
   (A) Cross pollinated crops
   (B) Self pollinated crops
   (C) Vegetatively propagated crops
   (D) Apomictic crop plants
8. The theory of 'Natural Selection' was put forward in
   (A) 1850 (B) 1859
   (C) 1869 (D) 1865
9. The 'pure line' theory was proposed by :
   (A) Johannsen (B) Yule
   (C) Bateson (D) Nilsson Ehle
10. Wheat varieties NP4 and NP6 are developed through
    (A) Mass selection
    (B) Heterosis breeding
    (C) Pure line selection
    (D) Pedigree breeding

11. Pure line breed refers to

(A) Heterozygosity only
(B) Homozygostiy only
(C) Homozygosity and self assortment
(D) Heterozygosity and linkage

12. Johansen proposed

(A) Theory of heterosis
(B) Pure line theory
(C) Male sterility
(D) Hybridization

13. Pureline selection is

(A) A method of accumulating favourable alleles
(B) Used to improve cross-pollinated crops
(C) Used to improve self-pollinated crops
(D) Used in vegetatively propagated crops

14. In mass selection, plants are selected on the basis of :

(A) Phenotypes (B) Genotypes
(C) Homozygosity (D) Heterozygosity

15. In ------------ a large number of plants of similar phenotype are selected and their seeds are mixed together to constitute the new variety.

(A) Pureline Selection
(B) Mass Selection
(C) Pedigree Selection
(D) Bulk Method

16. Bulk population breeding is suitable for

(A) Fruit crops (B) Vegetable
(C) Small grains (D) Flower crop

17. Which is a common method used in cross-pollinated crops?

(A) Mass selection (B) Bulk method
(C) Pure line method (D) Pedigree method

18. In mass selection, plants are selected on the basis of

(A) Phenotypes (B) Genotypes
(C) Homozygosity (D) Heterozygosity

19. Which of the following breeding procedures are not used for producing new varieties of self-pollinated crop?

(A) Recurrent selection
(B) Mass selection
(C) Pure line selection
(D) Hybridization

20. When a group of phenotypically similar appearing plant is selected and harvested and their seeds are bulked, the process is known as

(A) Pedigree breeding
(B) Mass selection
(C) Bulk method of breeding
(D) Pure line selection

21. What do you mean by evaluation of plant growth based on the performance of their progenies?

(A) Phenol test (B) Progeny test
(C) TZ test (D) Pedigree test

22. Isolation of desirable plant types from a population is known as -----------

(A) Selection (B) Collection
(C) Conservation (D) Hybridisation

23. Which one of the following is a correct answer for pureline selection

(A) Single heterozygous and cross pollinated crop
(B) Single heterozygous and often cross pollinated crop
(C) Single homozygous and self pollinated crop
(D) Single homozygous and often cross pollinated crop

24. In which type of crop, the progeny test method is used for breeding purpose

(A) Cross pollinated crop
(B) Often cross pollinated crop
(C) Self incompatible crop
(D) Self pollinated crop

25. Which one of the following is incorrect answer with respect to pure line selection

(A) Achieves maximum improvement
(B) Have wide adoptation
(C) Extreme uniformity
(D) Easy identification

26. Which one of the following statement is correct with respect to mass selection?

(A) Select large number of plants with similar phenotype and mix their seeds
(B) Select minimum number of plants with similar phenotype and mix their seeds
(C) Select small number of plants with similar phenotype and mix their seeds
(D) Select very small number of plants with similar phenotype and mix their seeds

27. Which one of the following reason is not correct with respect to mass selection?

(A) widely adopted varieties
(B) No need for extensive yield trails
(C) Restrains considerable genetic variability
(D) Less improvement than pure line selection

28. Which one of the following statement is not relevant to pure line selection?

(A) Narrow adaptation
(B) Difficult to identify
(C) Easily identified
(D) Highly uniform

29. Which one of the following is correct with respect to mass selection?

(A) Narrow adaptation
(B) Difficult to identify
(C) Easily identified
(D) Highly uniform

ANSWER SHEET

1. **B**	2. **B**	3. **B**	4. **D**	5. **C**	6. **C**	7. **A**	8. **B**	9. **A**	10. **C**
11. **B**	12. **B**	13. **C**	14. **A**	15. **B**	16. **C**	17. **A**	18. **A**	19. **B**	20. **B**
21. **B**	22. **A**	23. **C**	24. **D**	25. **B**	26. **A**	27. **D**	28. **B**	29. **B**	

4.2. HYBRIDISATION

1. Despite conspicuous deterioration of maize upon continuous selfing, it is more tolerant to inbreeding than :

(A) Lima bean (B) Barley
(C) Buck wheat (D) Alfalfa

2. Cross of F_1 hybrid with either parents is called :

(A) Test cross (B) Backcross
(C) Single cross (D) Double cross

3. Cross between cytoplasmic male sterile x cytoplasmic male fertile :

(A) Produce sterile F_1
(B) Produce fertile F_1
(C) Produce fertile, sterile in 1 : 1 ratio
(D) Produce fertile, sterile in 3 : 1 ratio

4. Composites are developed by using :

(A) Recurrent selection for specific combining ability
(B) Hybrid directly from intervarietal crosses
(C) Advanced generation seed mixtures obtained from high yielding intervarietal or interracial crosses
(D) The hybrid of an inbred with an open-pollinated variety

5. Evaluation of inbreds in maize is carried out in four stages. Open pollinated variety is involved in the :

(A) Single cross evaluation
(B) Double cross evaluation
(C) Phenotypic evaluation
(D) Top cross evaluation

6. Test cross involves the principle that :

(A) Test cross progeny directly corresponds with the types produced
(B) Test cross progeny directly corresponds with the type of gametes produced by F_1
(C) Test cross progeny directly corresponds with the type and proportions of gametes in F_1
(D) Test cross progeny directly corresponds with the type and proportions of gametes produced in F_2

7. After emasculation, the spike of the male parent in wheat :

(A) Is covered with a butter paper bag
(B) Need not be covered
(C) Crossed with selected male parent pollen
(D) Crossed and covered

8. The increased performance of F_1 hybrid over its parents is called :

(A) Dominance (B) Crossing over
(C) Heterosis (D) Over dominance

9. Emasculation is :

(A) Removal of male and female parts
(B) Removal of male parts from a bisexual flower
(C) Removal of female parts
(D) Removal of flower

10. In India male sterility is used in :

(A) Wheat (B) Maize
(C) Pearl millet (D) Chickpea

11. In India for hybrid seed production in maize we use :

(A) Cytoplasmic genetic male sterility
(B) Genetic male sterility
(C) Detasseling
(D) Self incompatibility

12. In cotton, the popularly grown commercial hybrid 'varalaxmi' is the product of :

(A) Intraspecific hybridization in *Gossypium hirsutum*
(B) Intraspecific hybridization in *Gossypium barbadense*
(C) Interspecific hybridization between *Gossypium arboretum* and *Gossypium herbaceum*
(D) Interspecific hybridization between *Gossypium hirsutum* and *Gossypium barbadense*

13. In bajra :

(A) Self incompatibility is found
(B) Cytoplasmic male sterility is found
(C) Genetic male sterility is found
(D) Cytoplasmic genetic male sterility is found

14. In ------------, the first studies on artificial hybridization for which yields were reported were those of Beal during the period between 1877 and 1882.

(A) Rye (B) Sorghum
(C) Maize (D) Pearl millet

15. In ------------ pollen dispersal is poor and seed set is likely to be low on male sterile plants.

(A) Tomato (B) Carrot
(C) Broccoli (D) Cauliflower

16. Restorer gene is needed :

(A) To make a plant correct
(B) To enhance the yield
(C) To make a plant resistant
(D) To bring fertility in male sterile plants

17. Selfing of a plant increases :

(A) Homozygosity (B) Heterozygosity
(C) Homogenesity (D) Heterogenesity

18. Specific combining ability of the parents should be high in order to develop :

(A) Synthetics (B) Varieties
(C) Composites (D) Hybrids

19. The 1st maize hybrid developed in India was :

(A) Vikram (B) Vijay
(C) Ganga - 2 (D) Kissan

20. High performance is noted in hybrid coconuts of :

(A) Tall x Dwarf (B) Dwarf x Tall
(C) Tall x Tall (D) Dwarf x Dwarf

21. Double cross in maize was done first by :

(A) Beadle (B) G.H. Shull
(C) D.F. Jones (D) Comstock

22. Test cross is back cross to :

(A) Dominant parent
(B) Recurrent parent
(C) Recessive parent
(D) Donor parent

23. Hybrid seeds can be produced through open pollinated with the use of :

(A) Clones (B) Inbreds
(C) Pure lines (D) Male sterile lines

24. Heterosis is commercially exploited in :

(A) Self pollinated species
(B) Cross pollinated species
(C) Often cross pollinated species
(D) Vegetatively propagated species

25. The first inbreeding experiments with maize were those reported by :

(A) Sprengel in 1866
(B) Darwin in 1876
(C) Sprague in 1886
(D) Baird in 1896

26. Polycross technique is adopted in improvement of :

(A) Maize (B) Bajra
(C) Guinea grass (D) Sunflower

27. By vaccum emasculation number of spikelets emasculated per day in paddy is :

(A) 1000 (B) 600
(C) 1500 (D) 100

28. Ganga safed is a hybrid of :

(A) Maize (B) Sorghum
(C) Cotton (D) Groundnut

29. A maintainer line is :

(A) Male fertile and can restore fertility
(B) Male sterile but cann't restore fertility
(C) Male sterile but can restore fertility
(D) Male fertile but cann't restore fertility

30. A hybrid is always :

(A) Homozygous (B) Heterozygous
(C) Both of the above
(D) Often homozygous

31. A homozygous line developed by self fertilization in a cross-pollinated crop is termed as :

(A) Mutant (B) Clone
(C) Inbred (D) Pureline

32. The performance of a double cross hybrid (A x C) x (B x D)can be predicted from the average performance of the combination of:

(A) A x B, A x D, B x C and C x D
(B) A x B, A x C, C x B and C x D
(C) A x B, A x C, A x D and B x C
(D) A x B, C x D, A x C and B x D

33. In cross pollinated species, a true breeding line obtained by continuous inbreeding is called :

(A) Synthetic variety
(B) Composite variety
(C) Clone (D) Inbred

34. The method(s) used for fixation of heterosis are :

(A) Apomixis

(B) sexual reproduction

(C) Monoploid

(D) Vegetative propagation

35. The dominance hypothesis of heterosis was first proposed by :

(A) Davenport (1908)

(B) East (1908)

(C) Hull (1945)

(D) Shull (1914)

36. CO 43 paddy variety was evolved by crossing :

(A) ADT 5 x IR 50 (B) ASD 5 x IR 20

(C) Dasal x IR 20 (D) Dasal x IR 8

37. Maintenance and multiplication of 'A' line used in crops like pearl millet, is done through :

(A) Crossing the plants of 'A' line with those of 'R' line

(B) Selfing the plants of 'A' line

(C) Crossing the plants of 'A' line with those of 'B' line

(D) Crossing the plants of 'A' line with any popular variety

38. Which one of the following is the simplest method for testing the General Combining Ability of an inbred line?

(A) Top cross (B) Test cross

(C) Single cross (D) Multiple cross

39. Which system is used for hybrid seed production in Onion?

(A) Self-incompatibility

(B) Dioecious

(C) Genetic male sterility

(D) Cytoplasmic male sterility

40. The number of three way crosses are :-

(A) n(n + 1)(n + 2)/2

(B) n(n - 1)(n - 2)/2

(C) n(n - 1)(n - 2)(n - 3)/8

(D) None of these

41. In maize, cross pollination occurs due to :

(A) Protandry

(B) Protogyny

(C) Self incompatibility

(D) Self sterility

42. Production of a hybrid in sesamum remains a constrain due to :

(A) Non-availability of genic male sterile lines

(B) High cost of seed production

(C) Both (A) and (B)

(D) Incompatibility reaction among varieties

43. Which one of the following statements regarding selfing is correct?

(A) It improves heterozygosity

(B) It does not have any effect on heterozygosity

(C) It decreases homozygosity

(D) It reduces heterozygosity

44. First cotton hybrid variety in the world developed in India, the variety is :

(A) Varalaxmi

(B) H4

(C) Suguna

(D) None of these

45. The component lines used in development of composite variety of maize are :

(A) Inbred lines

(B) Open pollinated varieties

(C) Single crosses

(D) Top crosses

46. Inbred lines in maize are maintained by:

(A) Self pollination

(B) Cross pollination

(C) Open pollination

(D) Sib pollination

47. Superiority of F_1 hybrid over a standard commercial variety is called :

(A) Heterobeltiosis
(B) Relative heterosis
(C) Economic heterosis
(D) Luxuriance

48. The factors responsible for cytoplasmic male sterility in plants are located in the:

(A) RNA
(B) Nuclear DNA
(C) Chloroplast DNA
(D) Mitochondrial DNA

49. ------------ deteriorates drastically upon inbreeding.

(A) Okra (B) Brinjal
(C) Carrot (D) Broccoli

50. Interaction of genes of inhibition of chlorophyll expression in a dihybrid results in a F_2 phenotypic ratio of :

(A) 9:7 (B) 13:3
(C) 15:1 (D) 10:6

51. Hybridization between species followed by polyploidy or chromosome doubling is known as :

(A) Allopolyploid
(B) Aneuploid
(C) Species differentiation
(D) Autopolyploid

52. A and B lines behave differently because :

(A) Genetically dissimilar
(B) Genetically similar for all traits
(C) They are isogenic for sterility
(D) Derived from different genetic sources

53. Emasculation of flower is necessary for hybridization in :

(A) Male sterile plants
(B) Self pollinated plants
(C) Self incompatible plants
(D) Dioecious plants

54. Top cross is used to find out :

(A) Combining ability of inbreds
(B) Specific combining ability of hybrid combinations
(C) General combining ability of inbreds
(D) Both GCA and SCA

55. Double cross plan was suggested by :

(A) Knight (B) Shull
(C) Jones (D) Beal

56. Systematic hybridization on a scientific basis began with the work :

(A) Koelreuter - *Nicotiana*
(B) Patel - Cotton
(C) Shull - Maize (D) Knight - Fruits

57. The cross of an inbred line with an open pollinated variety is known as:

(A) Double cross (B) Three way cross
(C) Top cross (D) Backcross

58. An example of a heterozygous but homogeneous population is :

(A) Pure line (B) Hybrid variety
(C) Open pollinated variety
(D) Synthetic variety

59. Cytoplasmic male sterility is best suited for hybrid seed production in :

(A) SP crops (B) CP crops
(C) Vegetatively propagated crops
(D) All above

60. A single hybrid of corn is pollinated with another line to produce :

(A) A double hybrid
(B) Three way cross
(C) An inbred variety
(D) Test cross variety

61. The term cybrid refers to the :

(A) Gametic hybrids
(B) Somatic hybrids
(C) Both (A) and (B)
(D) None of these

62. A F_1 plant with the genetic constitution AaBbccDd is test crossed with aabbccdd, there will be :

(A) 4 genotypes (B) 8 genotypes
(C) 10 genotypes (D) 12 genotypes

63. A mating among the following can not lead to heterosis :

(A) AAbb x aaBB
(B) aaBBcc x AABBCC
(C) AABBccdd x aabbCCDD
(D) AABBccdd x AABBccdd

64. Sterility is most pronounced in the F_1 generation following :

(A) Intervarietal hybridization
(B) Intravarietal hybridization
(C) Interspecific hybridization
(D) Intraspecific hybridization

65. First hybrid in pulses (Arhar) in India, developed through genetic male sterility :

(A) IPCH-8 (B) IHPC-8
(C) IHCP-8 (D) ICPH-8

66. Harvested seed of a hybrid cannot be used to raise the next crop because :

(A) Genetic purity is lost
(B) Hybrid vigour is lost generation after generation
(C) Recombination of genes occur fastly
(D) Degeneration of seed qualities take place

67. Inability of a hybrid to produce viable offspring is known as :

(A) Hybrid sterility
(B) Hybrid inviability
(C) Hybrid breakdown
(D) All of these

68. Term heterosis was introduced by :

(A) Hull (1945) (B) Shull (1914)
(C) Davenport (1908) (D) East (1936)

69. Highest uniformity is present in :

(A) Single cross
(B) Double cross
(C) Three way cross
(D) Multiple cross

70. Double top cross refers to :

(A) (A x B) x (A x B)
(B) (A x B) x (C x D)
(C) (A x B) x C
(D) (A x B) x OPV

71. ------------ is a reproductive deficiency of same plants where male organs in hermaphrodite flowers are rendered defunction

(A) Self incompatibility
(B) Male Sterility
(C) Cleistogamy
(D) Chasmogamy

72. The superiority of F_1 over its parents is known as

(A) Composite (B) Heterosis
(B) Cybirds
(D) Inbreeding depression

73. Cross of F1 hybrid with either parent is called

(A) Test cross (B) Single cross
(C) Back cross (D) None of them

74. The first inbreeding experiments with maize were those reported by

(A) Sprengel in 1866
(B) Darwin in 1876
(C) Sprague in 1886
(D) Baird 1896

75. The segregation of individuals in the F_2 or in a later generation of a cross, which shows a more extreme development of a character than either parent is termed as

(A) Hybridisation
(B) Heterosis
(C) Linkage
(D) Transgressive segregates

76. An example of a heterozygous but homogenous population is

(A) Pure line
(B) Hybrid variety
(C) Synthetic variety
(D) Open pollinated variety

77. A mating among the following can not lead to heterosis

(A) AAbb x aaBB
(B) AABBccdd x AABBccdd
(C) AABBccdd x aabbCCDD
(D) aaBBcc x AABBCC

78. Heterosis is

(A) Appearance of spontaneous mutations
(B) Induction of mutations
(C) Mixture of two or more traits
(D) Superiority of hybrids over their parents

79. The new varieties of plants are produced by

(A) Introduction and mutation
(B) Selection and hybridisation
(C) Mutation and selection
(D) Selection and introduction

80. Bagging is done to

(A) Avoid cross pollination
(B) Avoid self pollination
(C) Achieve desired pollination
(D) Prevent contamination from foreign pollen

81. Mating between two individuals related by descent is

(A) Selfing (B) Inbreeding
(C) Outbreeding
(D) Maintenance of population

82. Mating between a inbred line and an open pollinated variety is called

(A)Top cross
(B) Poly cross
(C) Double top cross
(D) Composite cross

83. Superiority of F_1 over the better is called

(A) Heterobeltiosis
(B) Average Heterosis
(C) Negative Heterosis
(D) Economic Heterosis

84. Production of hybrid plants through the fusion of protoplasts of two different plant species / varieties is called

(A) Distant hybridization
(B) Somatic hybridization
(C) In vitro pollination
(D) Wide hybridization

85. What do you mean by hybridization?

(A) Mating of homozygous genotypes
(B) Mating of homogenous genotypes
(C) Mating of dissimilar genotypes
(D) Mating of similar genotypes

86. Which gene is involved in transfer of one or more characters into a single variety from other variety?

(A) Oligogenes (B) Polygenes
(C) Both A & B (D) Single gene

87. In which group of crops, hybridization is recommended for breeding?

(A) Self pollinated crops
(B) Cross pollinated crops
(C) Often cross pollinated crops
(D) Often self pollinated crops

88. Which one of the following is a single cross hybrid?

(A) AxR --- F1 (B) AxB--- F1
(C) AxA----F1 (D) BxR----F1

89. If more than one parents are involved in hybrid development, that hybrid is known as -------------

(A) Single cross (B) Test cross
(C) F1 hybrid (D) Convergent cross

90. What is the parent of triticale?

(A) *Triticum* sp X *Secale cereale*

(B) *Triticum* sp X *Oryza sativa*

(C) *Triticum* sp X *Sorghum vulgare*

(D) *Triticum* sp X *Hordeum* sp

91. In rice CO 31 variety is developed by using ------------- as parental lines

(A) *Oryza sativa* var.*japonica* X *Oryza perennis*

(B) *Oryza sativa* var.*indica* X *Oryza perennis*

(C) *Oryza sativa* var.*asiatica* X *Oryza perennis*

(D) *Oryza sativa* var.*asiatica* X *Oryza officinalis*

92. Parents involved in hybridization between two strains of same species is called as -----

(A) Intervarietal hybridization

(B) Distant hybridization

(C) Interspecific hybridization

(D) Intergeneric hybridization

93. Which one of the following is a correct answer for a hybrid obtained from crossing between two different species of the same genera?

(A) Intervarietal hybridization

(B) Distant hybridization

(C) Interspecific hybridization

(D) Intergeneric hybridization

94. Which one of the following is a correct answer for a hybrid obtained from crossing between two different species of the different genera?

(A) Intervarietal hybridization

(B) Distant hybridization

(C) Interspecific hybridization

(D) Intergeneric hybridization

95. If a hybrid is developed by mating between two species of the same genus

(A) Intervarietal hybridization

(B) Distant hybridization

(C) Interspecific hybridization

(D) Intergeneric hybridization

96. ------------- is a hybrid between mating of two different genus

(A) Intervarietal hybridization

(B) Distant hybridization

(C) Interspecific hybridization

(D) Intergeneric hybridization

97. Which one of the following method will kill the pollen grains for hybridization?

(A) Hand emasculation

(B) Suction method

(C) Detasseling

(D) Hot water emasculation

98. Hand emasculation is used for ----------

(A) Cotton (B) Maize

(C) Rice (D) Sorghum

99. At what temperature, the hot water emasculation is done for sorghum hybridization?

(A) 41-42°C (B) 42-48°C

(C) 40-43°C (D) 41-47°C

100. A temperature of 40-44°C is maintained to kill pollen grains in ------------ hybridization

(A) Wheat (B) Sorghum

(C) Rice (D) Triticale

101. A cold water treatment @ 0-6°C is used for --------- hybridization

(A) Wheat (B) Sorghum

(C) Rice (D) Triticale

102. Which one of the following method is followed in maize hybridization program?

(A) Hand emasculation

(B) Detasseling

(C) Cold water treatment

(D) Hot water treatment

103. In which crop, detasseling is used for hybridization program?

(A) Cotton (B) Redgram
(C) Bajra (D) Maize

104. In which double cross hybrid is developed?

(A) Cotton (B) Redgram
(C) Bajra (D) Maize

105. In which crop, CMS is used fro hybridization?

(A) Cotton (B) Maize
(C) Rice (D) Redgram

106. In which crop, GMS is used fro hybridization?

(A) Cotton (B) Maize
(C) Rice (D) Tomato

107. In hybridization, which one of the following is a sterile line?

(A) B line (B) A line
(C) R line (D) AXR line

108. ---------line is a maintainer line which is used to maintain sterile line

(A) B line (B) A line
(C) R line (D) AXR line

109. Which one of the following is a restorer line in hybidization?

(A) B line (B) A line
(C) R line (D) AXR line

110. Which one of the following is a isogeneic sterile line in hybridization?

(A) B line (B) A line
(C) R line (D) AXR line

111. Which one of the following is incorrect with respect to two line breeding?

(A) Physical or manual method
(B) Use of GMS
(C) Use of CMS
(D) Use of EGMS

112. Which one of the following is a correct answer for three line breeding?

(A) Physical or manual method
(B) Use of GMS
(C) Use of CMS
(D) Use of EGMS

113. Which one of the following method is used fro Redgram hybridization ?

(A) Hand emascukation
(B) Use of GMS
(C) Use of CMS
(D) Use of EGMS

114. In which crop, single line breeding is followed to develop hybrids?

(A) Sorghum (B) Bajra
(C) Napier grass (D) Rice

115. Which gene controls the male sterility of plants?

(A) MsMs (B) MSMS
(C) MSms (D) msms

116. ------ is a dominant gene which controls male fertility?

(A) MsMs (B) MSMS
(C) MSms (D) msms

117. Which chemical is used chemical as a hybridizing agent?

(A) EMS
(B) Ethyl alcohol
(C) MH
(D) Methanol

118. In which crop, TGMS is used in hybridization program?

(A) Rice (B) Maize
(C) Sorghum (D) Bajra

119. In ------------, PGMS is followed in hybridization?

(A) Rice (B) Maize
(C) Sorghum (D) Bajra

120. Which one of the following method is suited for rice hybridization?

(A) Single line breeding
(B) Double cross method
(C) Triple cross method
(D) EGMS

121. During TGMS of rice hybridization, day and night temperature of ---------- is maintained to produce sterility

(A) 24 & 18°C (B) 32 & 24°C
(C) 20 & 16°C (D) 30 & 20°C

122. In PGMS, the plant becomes fertile if the plants are exposed to --------- day length

(A) > 14 h (B) < 14 h
(C) < 13 h (D) 12-14 h

123. Which one of the following hybrid is developed by intra specific hybridization?

(A) Varalakshmi (B) CSH 5
(C) CO H1 (D) Deccan

124. --------------- is a hybrid developed by inter specific hybridization?

(A) Varaiakshmi (B) CSH 5
(C) CO H1 (D) Deccan

125. Which one of the following is a single cross hybrid?

(A) Varalakshmi (B) CSH 5
(C) CO H1 (D) Deccan

126. Which one of the following hybrid is developed by double cross hybridization?

(A) Varalakshmi (B) CSH 5
(C) CO H1 (D) Deccan

127. Which one of the following is a three way cross hybrid?

(A) Varalakshmi (B) CSH 5
(C) CO H1 (D) Deccan

ANSWER SHEET

1. D	2. B	3. A	4. C	5. D	6. C	7. B	8. C	9. B	10. C
11. C	12. D	13. D	14. C	15. A	16. D	17. A	18. D	19. C	20. A
21. C	22. C	23. B	24. B	25. B	26. C	27. B	28. A	29. D	30. B
31. C	32. A	33. D	34. A	35. A	36. C	37. C	38. B	39. D	40. B
41. A	42. C	43. B	44. B	45. B	46. A	47. C	48. D	49. C	50. B
51. A	52. C	53. B	54. C	55. C	56. A	57. C	58. B	59. C	60. B
61. B	62. B	63. D	64. C	65. D	66. B	67. A	68. B	69. A	70. D
71. D	72. B	73. C	74. B	75. D	76. B	77. B	78. D	79. B	80. D
81. C	82. A	83. A	84. B	85. C	86. C	87. A	88. B	89. D	90. A
91. B	92. A	93. B	94. B	95. C	96. D	97. D	98. A	99. B	100. C
101. C	102. B	103. D	104. D	105. C	106. A	107. B	108. A	109. C	110. A
111. C	112. C	113. BC	114. C	115. B	116. C	117. A	118. A	119. A	120. D
121. C	122. C	123. B	124. A	125. C	126. D	127. D			

4.3. PEDIGREE METHOD/BULK METHOD/MULTILINE VARIETIES/ BACKCROSS METHOD

1. In pedigree breeding single plants are selected in which one of the following generation :

(A) F_2 (B) F_3
(C) F_4 (D) F_5

2. A method of breeding most commonly used when the desired variation is required to be induced in a vegetatively propagated crop is :

(A) Pedigree method of breeding
(B) Mutation breeding
(C) Polyploidy
(D) Backcross method of breeding

3. Swarna a variety of sorghum is a :

(A) Single cross hybrid
(B) Double cross hybrid
(C) Composite
(D) Variety developed through pedigree method

4. Mass pedigree selection method of breeding is particularly, useful for improvement of :

(A) Self incompatible crops
(B) Self compatible cross pollinated crops
(C) Self pollinated crops
(D) Dioecious crops

5. A single desirable plant can :

(A) Be used in pedigree selection
(B) Be used to produce a variety in cross pollinated crop
(C) Be used in synthetic variety
(D) Be used in bulk method of breeding

6. In pedigree breeding single plants are selected in which one of the following generation

(A) F_5 (B) F_3
(C) F_2 (D) F_1

7. Method of breeding can be used for studying the ancestral characteristic

(A) Pure line (B) Pedigree method
(C) Mass selection (D) Hybridisation

8. For what purpose the pedigree method is used?

(A) To develop hybrid variety
(B) To develop pureline variety
(C) To develop resistant variety
(D) To develop mutant variety

9. Which one of the following is not relevant to pedigree method?

(A) Efficient use of breeders skill and judgement in selection
(B) For improving characters
(C) Take less time
(D) Take more time

10. Bulk method is otherwise called as ---------

(A) Pedigree method
(B) Selection method
(C) Mass method
(D) Single seed descent method

11. What do you mean by backcross method?

(A) Cross between F1 and standard variety
(B) Cross between male parent and restorer parent
(C) Cross between F1 and a male parent
(D) Cross between female parent and male parent

12. For what purpose, the back cross method is recommended?

(A) To develop a resistance mechanism
(B) To develop a high yielding variety
(C) To develop a mutant variety
(D) To develop a hybrid variety

13. Which one of the following is a correct answer for multiline variety?

(A) Bulk population of parental lines after open pollination
(B) Mixture of intervarietal cross
(C) Mixture of several purelines
(D) Mixture of hybrid and its parents

14. How many backcrosses are done in backcross method?

(A) 5 crosses (B) 4 crosses
(C) 6 crosses (D) 3 crosses

15. How many backcrosses are usually carried out in restricted backcross method?

(A) 6 crosses (B) 7 crosses
(C) 5 crosses (D) 3 crosses

16. Which one of the following is a correct statement with respect to population approach method?

(A) Accumulation of desirable genes
(B) Susceptible to the pathogen
(C) New race may attack the line
(D) Limited recombination

17. Which one of the following is a correct answer for synthetic variety?

(A) Bulk population of parental lines after open pollination
(B) Mixture of intervarietal cross
(C) Mixture of several purelines
(D) Mixture of hybrid and its parents

18. Which one of the following statement is not relevant to synthetic variety?

(A) Crop species where pollination is difficult
(B) More genetic variability
(C) Performance is lower than a hybrid
(D) Greater opportunity for recombination

19. Which one of the following answer is correct with regard to composites?

(A) Bulk population of parental lines after open pollination
(B) Mixture of intervarietal cross
(C) Mixture of several purelines
(D) Mixture of hybrid and its parents

20. In backcross breeding repeated crossing with recurrent parent is done to :

(A) Break undesirable linkage
(B) Transfer the desirable gene from donor parent
(C) Completely recover the genotype of recurrent parent
(D) Retain the intensity of desirable gene

21. What is the most effective method for transfer of on oligogenic character in an otherwise agronomically superior genotype?

(A) Backcross
(B) Bulk population
(C) Pedigree
(D) Recurrent selection

ANSWER SHEET

1. A	2. C	3. D	4. C	5. A	6. C	7. B	8. B	9. D	10. C
11. C	12. A	13. C	14. C	15. D	16. A	17. A	18. D	19. B	20. C
21. A									

Chapter - 5

BREEDING METHODS FOR CROSS POLLINATED CROPS

5.1. HETEROSIS AND INBREEDING

1. What is the incorrect answer with respect to effects of inbreeding?
   (A) Appearance of lethal and sublethal alleles
   (B) Reduction in vigour
   (C) Time consuming
   (D) Increase in homozygosity

2. Whoch one of the following is a correct answer with respect to inbreeding?
   (A) Time consuming
   (B) Inbreeding depression
   (C) Poor performance
   (D) Identical to recurrent parent

3. Which one of the manifestations is incorrect with regard to heterosis?
   (A) Increased opportunity for recombination
   (B) Increased yield
   (C) Increased reproductive ability
   (D) Increased resistance to pests and diseases

4. What is the effect of dominance hypothesis?
   (A) Unfavourable effect
   (B) Harmful effect
   (C) Deleterious effect
   (D) Favourable effect

ANSWER SHEET

1	**C**
2	**B**
3	**A**
4	**D**

Chapter - 6

BREEDING METHODS FOR ASEXUALLY REPRODUCED CROPS

6.1. CLONAL SELECTION

1. Which one of the following answer is a incorrect answer with regard to asexually reproduced crop?

(A) Mostly monoploids

(B) Mostly perennials

(C) Show reduced flowering and seed set

(D) Mostly cross pollinated

2. Which one of the following statement is correct pertaining to clones?

(A) Identical individuals

(B) No phenotypic variations

(C) No phenotype X environment interactions

(D) Intra specific hybrids

3. --------- is most benefited by inter specific hybridization

(A) Ginger (B) Turmeric

(C) Potato (D) Tapioca

4. Which one of the following potato variety was developed from a complex cross (*Solanum curtilobum X S. tuberosum) X S. andigena*

(A) Kufri muthu (B) Kufri kuber

(C) Kufri red (D) Darjeeling red

5. Which one of the following answer is not correct with regard to the effect of clonal selection?

(A) Genotype effect

(B) Environment effect

(C) GXE interaction effect

(D) PXE interaction effect

ANSWER SHEET

1. **A** 2. **A** 3. **C** 4. **B** 5. **D**

Chapter - 7

RESISTANCE BREEDING

1. The gene-for-gene hypothesis regarding disease resistance in crop varieties has following features :

   1. This hypothesis was proposed for rust in linseed by Flor.
   2. For each gene for resistance in the host plant, there is corresponding gene for virulence in the pathogen.

   Which of the features given above is/are correct?

   (A) 1 only (B) 2 only
   (C) Both 1 and 2 (D) Neither 1 nor 2

2. Breeding for resistance is important in urd or mung for :

   (A) Yellow Mosaic Virus
   (B) Damping off
   (C) Leaf spot
   (D) Cercospora leaf spot

3. Breeding method for transferring monogenic disease resistance from wild species to improved variety of wheat is :

   (A) Bulk method
   (B) Backcross method
   (C) Pedigree method
   (D) Pure line selection

4. Which of the following is not a drought resistant donor in rice :

   (A) PTB 29 (B) Oorpandy
   (C) TKM 1 (D) PTB 30

5. The concept of vertical and horizontal resistance was introduced by :

   (A) Robinson (1971)
   (B) Nelson (1973)
   (C) Vander Plank (1963)
   (D) Russell (1978)

6. An amino acid which is related to drought tolerance in plants is :

   (A) Glutamic acid (B) Proline
   (C) Glycine (D) Hydroxyproline

7. Transfer of resistance gene to high yielding self-pollinated varieties is achieved by

   (A) Hybridisation and pedigree selection
   (B) Recurrent backcrossing and selection
   (C) Hybridisation and bulk selection
   (D) Hybridisation and mass selection

8. Biffin in 1905 announced, that resistance to yellow rust of wheat is governed by -------- gene in crosses

   (A) Single dominant
   (B) Single recessive
   (C) Double dominant
   (D) Double recessive

9. In rice, resistance to stem bore is associated with:

   (A) High tannin in stem
   (B) High silica content in shoot
   (C) Waxiness of leaves
   (D) High aspartic acid in leaves

10. Under drought conditions. There is an increase in:

(A) Abscisic acid content
(B) Ethylene level
(C) Proline level
(D) NAA level

11. Which one of the following variety of potato is resistant to YVMV?

(A) Darjeeling red (B) Kufri red
(C) Kufri muthu (D) Kufri kuber

12. Which one if the following variety of bhendi is resistant to YVMV?

(A) Parbhani kranti
(B) varsha uphar
(C) Pusa sawani
(D) Arka anamika

13. In cotton ------------ is resistant to red leaf blight

(A) Laxmi (B) Gadagl
(C) Varalaxmi (D) MCU5

14. Which one of the following breeding method is used to transfer disease resistance from one variety to other variety?

(A) Mass selection
(B) Hybridization
(C) Pure line selection
(D) Pedigree method

15. Resistance to ------ was developed by selection.

(A) Cotton leaf hopper
(B) Chilli leaf hopper
(C) Tomato leaf hopper
(D) Potato leaf hopper

16. How the genes for drought resistance are transferred to high yield variety?

(A) Hybridization
(B) Mass selection
(C) Pure line selection
(D) Pedigree method

ANSWER SHEET

1. **C**	2. **A**	3. **B**	4. **B**	5. **C**	6. **B**	7. **B**	8. **B**	9. **B**	10. **C**
11. **B**	12. **C**	13. **A**	14. **B**	15. **D**	16. **A**				

Chapter 8

MUTATION BREEDING

1. An organism which has acquired a heritable variation as a result of sudden change in the hereditary material is called as a :

   (A) Albino (B) Biotype
   (C) Mutant (D) Hybrid

2. Sudden and sharp change in individual and is directly inherited by the offspring is called :

   (A) Adaptation
   (B) Heredity
   (C) Mutation
   (D) Continuous variation

3. Mutation is an altered phenotype and results in a change which is :

   (A) Temporarily inherited
   (B) Inherited
   (C) Partly inherited
   (D) Unstable

4. Hugo de Vries, who defined the 'mutation' was :

   (A) French Bacteriologist
   (B) Russian Biologist
   (C) English Botanist
   (D) Dutch Botanist

5. At the molecular level point mutations are due to :

   (A) Destruction of the double helix
   (B) Destruction of the bases
   (C) Alterations of sequences of base in DNA
   (D) Shifting of a portion of the helix and bases

6. Formation of new genes take place due to :

   (A) Mutation (B) Transduction
   (C) Transversion (D) Inversion

7. The word 'Mutation' was first used by :

   (A) Muller (B) S. Wright
   (C) Morgan (D) Hugo de Vries

8. Recommended concentration of EMS for induced mutagenesis is :

   (A) 1.00 to 3.5% (B) 0.03 to 1.5%
   (C) 0.30 to 3.0% (D) Above 4%

9. The equilibrium in random mating population is distributed by selection, mutation, inbreeding, migration and genetic drift in which situation the population may regain its original composition?

   (A) Mutation
   (B) Migration
   (C) Inbreeding and selection
   (D) Genetic drift

10. Which of the following rice variety has been developed through induced mutagenesis?

    (A) Jagannath (B) Pankaj
    (C) Mahsuri (D) IR 8

11. The chemical mutagens usually induce :
(A) Chromosomal breakages
(B) Gene mutation
(C) Chromosomal translocation
(D) None of these

12. Mutation due to change in gene because of:
(A) Addition (B) Loss
(C) Degeneration (D) All of these

13. Who discovered 'X-rays' first?
(A) Wilson (B) Muller
(C) Benzer (D) Roentgen

14. Which one of the following mutagens is non-ionizing types?
(A) Gamma rays (B) Beta rays
(C) X - rays (D) Ultra - violet rays

15. Reversion occurs due to :
(A) Mutation
(B) Gene interaction
(C) Gene interaction and mutation
(D) None of these

16. The quickest method of plant breeding is
(A) Introduction (B) Selection
(C) Hybridisation (D) Mutation Breeding

17. At the molecular level point mutations are due to
(A) Destruction of the bases
(B) Destruction of the double helix
(C) Shifting of a portion of the helix and bases
(D) Alterations of sequences of base in DNA

18. Formation of new genes takes place due to
(A) Inversion (B) Transduction
(C) Transversion (D) Mutation

19. Hugo de Vries, who defined the 'mutation', was
(A) Dutch Botanist
(B) French Bacteriologist
(C) English Botanist
(D) Russian Biologist

20. Most of the mutation have effect
(A) Deleterious (B) Good
(C) Bad (D) Both (B) & (C)

21. Mutation is known to be due to change in genes
(A) Loss (B) Gain
(C) Deletion (D) Additive

22. Para-mutation in ------------ is perhaps the one and only example of blending of alleles
(A) Sorghum (B) Maize
(C) Pearl millet (D) Foxtail millet

23. Sudden and sharp change in individual and is directly inherited by the offspring is called
(A) Mutation
(B) Adaptation
(C) Heredity
(D) Continuous variation

24. The term mutation was coined by :
(A) Morgen (B) De Vries
(C) Mullar (D) Stadler

25. Ehyle methane sulphonate is :
(A) Base analogue
(B) Acridine dye
(C) Alkylatind agent
(D) Hormone

26. Stadler first used X-rays for induction of mutation in 1928 in :
(A) Drosophila (B) Wheat
(C) Barley (D) Maize

27. Jagannath is mutant variety of :
(A) Wheat (B) Rice
(C) Cotton (D) Castor

28. Which of the following mutagens is less penetrating?

(A) X-rays (B) Gamma rays
(C) Beta rays (D) UV rays

29. Sharbati Sorona is mutant variety of :

(A) Wheat (B) Rice
(C) Cotton (D) Maize

30. The International Atomic Energy Agency (IAEA) is located in :

(A) USA (B) Australia
(C) Austria (D) Germany

31. The term mutation was introduced by ------ in 1900

(A) Hugo Milton (B) Hugo Davis
(C) Hugo de vries (D) Hugo Vozzo

32. In which year, the mutation term was coined by Hugo de vries

(A) 1800 (B) 1900
(C) 1809 (D) 1909

33. Which one of the following plant part is used for mutagen treatment?

(A) Ovule (B) Ovary
(C) Zygote (D) Seed

34. ---------- is the dosage of mutagen that produces maximum frequency of mutation and causes minimum killing

(A) LD 30 (B) LD 40
(C) LD 60 (D) LD 50

35. which one of the following is not relevant in case of demerits of mutation breeding?

(A) Creates inexhaustible variation
(B) Low frequency of desirable mutation
(C) Associated with undesirable effects
(D) Most of the mutants are recessive

36. ------------ is the classical chemical mutagen that is used in mutation breeding

(A) Sodium chloride
(B) Calcium chloride
(C) Ethyl Methyl sulphonate
(D) Ethyl Methyl carbonate

37. What is the effect of mutation at chromosome level?

(A) Transcription (B) Transversion
(C) Translation (D) Replication

38. Which one of the following is not relevant to mutational changes?

(A) Transcription (B) Transversion
(C) Translation (D) Replication

39. In which crop, Hugo de vries has noticed sudden heritable changes?

(A) *Oryza* (B) *Oleodendron*
(C) *Oenothera* (D) *Oxalis*

40. Which one of the following is not an ionizing radiation agent used in mutational breeding?

(A) Ultraviolet light
(B) Infrared light
(C) Gamma rays
(D) X-rays

41. EMS react with ------- by adding ethyl group which causes DNA replication machinery during the process of mutation.

(A) Guanine (B) Adenine
(C) Cytosine (D) Uracil

42. In which crop species, mutagenic effect was observed?

(A) Oats (B) Wheat
(C) Barley (D) Triticale

ANSWER SHEET

1. **C**	2. **C**	3. **B**	4. **D**	5. **C**	6. **A**	7. **D**	8. **B**	9. **A**	10. **A**
11. **B**	12. **D**	13. **D**	14. **D**	15. **D**	16. **D**	17. **D**	18. **D**	19. **A**	20. **A**
21. **A**	22. **B**	23. **A**	24. **B**	25. **C**	26. **C**	27. **B**	28. **D**	29. **A**	30. **C**
31. **C**	32. **B**	33. **D**	34. **A**	35. **A**	36. **C**	37. **B**	38. **D**	39. **C**	40. **B**
41. **A**	42. **C**								

Chapter 9

POLYPLOIDY BREEDING

1. Which one of the following is a polyploidy?

   (A) Haploid (B) Diploid

   (C) Monoploid (D) Triploid

2. In which polyploidy, chromosome doubling within a species take place?

   (A) Autopolyploidy

   (B) Allopolyploidy

   (C) Euploidy

   (D) Aneuploidy

3. Which one of the following crop is not a triploid?

   (A) Banana (B) Maize

   (C) Apple (D) Ginger

4. Which one of the following crop is a tetraploid?

   (A) Banana (B) Maize

   (C) Apple (D) Cotton

5. Which one of the following crop is not a hexaploid?

   (A) Maize

   (B) Oats

   (C) Chrysanthemum

   (D) Triticale

6. In polyploidy breeding,the dosage of colchicine for seed treatment is -------

   (A) 0.1 to 1%

   (B) 0.5 to 1%

   (C) 0.001 to 1%

   (D) 0.01 to 1%

7. For treatment of growing shoot apices, the colchicine is used at the concentration of -------- fro polyploidy breeding

   (A) 0.1 to 1%

   (B) 0.5 to 1%

   (C) 0.001 to 1%

   (D) 0.01 to 1%

ANSWER SHEET

1. **D** 2. **A** 3. **B** 4. **D** 5. **A** 6. **B** 7. **A**

Part 2
BIOTECHNOLOGY

Chapter 1

CELL BIOLOGY

1. During oogenesis the cell that is fertilized by a sperm to become an egg is
(A) Primary spermatocyte
(B) Secondary oocyte
(C) Primary oocyte
(D) Secondary spermatocyte

2. A bivalent consists of
(A) Two chromatids and one centromere
(B) Four chromatids and two centromeres
(C) Two chromatids and two centromeres
(D) Four chromatids and four centromeres

3. A change in the amount of yolk and its distribution in the egg will affect:
(A) Fertilization
(B) Formation of zygote
(C) Pattern of cleavage
(D) Number of blastomeres produced

4. Chromosomes can be counted best at the stage of
(A) Metaphase (B) Late anaphase
(C) Telophase (D) Late prophase

5. The stage in which daughter chromosomes move toward the poles of the spindle is
(A) Anaphase (B) Metaphase
(C) Prophase (D) Telophase

6. Chiasmata are first seen in
(A) Pachytene (B) Zygotene
(C) Leptotene (D) Diplotene

7. Protein subunit found within microtubules is
(A) Collagen (B) Tubulin
(C) Myosin (D) DNA

8. Balbiani rings occur in
(A) Autosome
(B) Heterosome
(C) Lampbrush chromosome
(D) Polytene chromosome

9. Cell plate grows from
(A) One wall to another
(B) Centre to wall
(C) Wall to centre
(D) Simultaneously

10. Which of the following cellular structures always disappears during mitosis and meiosis?
(A) Plasma membrane
(B) Nucleolus and nuclear envelope
(C) Plastids
(D) Cell membrane

11. Centromere is a constituent of
(A) Ribosome
(B) ER
(C) Chromosome
(D) Mitochondrion

12. Chiasma formation occurs in
(A) Pachytene (B) Zygotene
(C) Leptotene (D) Diplotene

13. Crossing-over results in
(A) Recombination of linked alleles
(B) Dominance of alleles
(C) Segregation of alleles
(D) Linkage between genes

14. Cytoplasmic division of a cell is called
(A) Cell plate formation
(B) Cytokinesis
(C) Mitosis
(D) Synapsis

15. DNA replication takes place during
(A) S - phase (B) G - phase
(C) G - phase (D) Prophase

16. During meiosis I, the number of chromosomes is
(A) Halved (B) Tripled
(C) Doubled (D) Quadrupled

17. During mitosis ER and nucleolus begin to disappear at
(A) Early prophase
(B) Late prophase
(C) Early metaphase
(D) Late metaphase

18. Crossing over occurs in
(A) Pachytene (B) Zygotene
(C) Leptotene (D) Diplotene

19. During which stage the chromosomes first become visible.
(A) Anaphase (B) Metaphase
(C) Prophase (D) Telophase

20. Exchange of paternal and maternal chromosome material during cell division is
(A) Crossing over
(B) Bivalent formation
(C) Dyad formation
(D) Synapsis

21. Haploid complement of chromosome of an organism is
(A) Genome (B) Phenotype
(C) Genotype (D) Genetic system

22. During gamete formation, the enzyme recombinase participates during
(A) Prophase-I (B) Prophase-II
(C) Metaphase-I (D) Anaphase-II

23. In Human secondary spermatocyte how many chromosomes are present?
(A) 45 (B) 23
(C) 30 (D) 15

24. Lampbrush chromosomes are observed in
(A) Mitotic prophase
(B) Mitotic metaphase
(C) Meiotic metaphase
(D) Meiosis prophase

25. In humans, at the end of the first meiotic division, the male germ cells differentiate into the:
(A) Spermatogonia
(B) Primary spermatocytes
(C) Secondary spermatocytes
(D) Spermatids

26. In mitosis, nuclear envelope and nucleolus disappear during
(A) Metaphase (B) Interphase
(C) Prophase (D) Telophase

27. Meiosis is best seen in
(A) Gamete
(B) Pollen grain
(C) Microsporangium
(D) Anther wall

28. Meiosis occurs for the human female in ______.
(A) Pancreas (B) Liver
(C) Ovary (D) Kidney

29. Cell was discovered by
(A) Leeuwenhoek
(B) Robert Hooke
(C) Robert Swanson
(D) Robert Brown

30. The spherical structured organelle that contains the genetic material is
(A) Cell walls (B) Ribosomes
(C) Nucleus (D) Mitochondria

31. Protoplasm found inside the nucleus is known as
(A) Amyloplast (B) Nucleoplasm
(C) cytoplasm (D) Elaioplast

32. Which of the following statements are true about Endoplasmic

Reticulum? (a) Smooth Endoplasmic Reticulum makes lipids. (b) It is also called the control center of the cell. (c) It processes carbohydrates. (d) It modifies chemicals that are toxic to the cell.
(A) (a), (b) and (c)
(B) (a), (c) and (d)
(C) only (a) and (d)
(D) all are correct

33. Prokaryotic genetic system has
(A) DNA but no histones
(B) Both DNA and histones
(C) Neither DNA nor histones
(D) Either DNA or histones

34. Which of the following statements are true about Eukaryotes? (A) They are cells with a nucleus. (B) They are found both in humans and multicellular organisms. (C) Endoplasmic reticulum is present in Eukaryotes. (D) They have chemically complexed cell wall.
(A) (A), (B) and (C)
(B) (A), (C) and (D)
(C) (A), (B) and (D)
(D) All are correct

35. Which one of the following organelles digests the old organelles that are no longer useful to the cells?
(A) Ribosomes (B) Mitochondria
(C) Lysosomes (D) Chromatin

36. Plasmodesmata are located in narrow areas of _______.
(A) Cell walls (B) Protoplasm
(C) Cellulose (D) Nuclei

37. What do prokaryotic cells lack?
(A) Cell membrane
(B) Cytoplasm
(C) Cell wall
(D) membrane-bound nucleus

38. Which of the following is an example of cell devoid of nuclear membrane and mitochondria is
(A) Bacterial cell (B) Sperm
(C) Protist (D) Sponge cell

39. Lampbrush chromosomes are seen in
(A) Prophase
(B) Mitotic metaphase
(C) Mitosis
(D) Meiotic prophase

40. Which one of the following is not considered as a part of the endomembrane system?
(A) Vacuole (B) Lysosome
(C) Golgi complex (D) Peroxisome

41. Animal cell differs from plant cells in possessing
(A) Plastid (B) Golgi body
(C) Vacuole (D) Centrosome

42. Carbohydrates present in the plasmalemma are in the form of
(A) Cellulose
(B) Hemicellulose
(C) Starch
(D) Glycolipids and glycoproteins

43. Cellular totipotency means

(A) Synthesis of new cells
(B) Formation of new plants
(C) Formation of new species
(D) Capability of a plant cell to form complete plant

44. Important site for formation of glycoproteins and glycolipids is

(A) Lysosome (B) Vacuole
(C) Golgi apparatus
(D) Plastid

45. In eubacteria, a cellular component that resembles eukaryotic cell is

(A) Cell wall
(B) Plasma membrane
(C) Nucleus
(D) Ribosomes

46. Intact chloroplast from green leaves can be isolated by:

(A) Acetone (B) Ethanol
(C) Sucrose solution
(D) Alcohol

47. Which of the following is considered an exception to cell theory?

(A) Protists (B) Mycoplasma
(C) Virus (D) Algae

48. Membrane-bound organelles are absent in

(A) Plasmodium (B) Saccharomyces
(C) Streptococcus (D) Chlamydomonas

49. Mitochondria do not occur in

(A) Bacteria (B) Green Algae
(C) Brown algae (D) Red Algae

50. The size of molecules, which can pass through the plasma membrane is

(A) 4 - 15 Å (B) 25 - 80 Å
(C) 8 - 10 Å (D) 10 - 35 Å

51. Plasmodesmata transports gibberellic acid in

(A) Protists (B) Mycoplasma
(C) Virus (D) Algae

52. Smaller cell is

(A) Less active metabolically
(B) With larger nucleus
(C) With smaller nucleus
(D) More active metabolically

53. The cytoplasmic connections from cell to cell are known as

(A) Middle lamella
(B) Plasmodesmata
(C) Cell membrane system
(D) Endoplasmic reticulum

54. The main difference between plant and animal cells is

(A) Plant cells has small vacuoles
(B) Animal cell lacks rigid cell wall
(C) Animal cells has large vacuoles
(D) Plant cell lacks rigid cell wall

55. The plasma membrane consists mainly of

(A) Proteins embedded in a carbohydrate bilayer
(B) Phospholipids embedded in a protein bilayer
(C) Proteins embedded in a phospholipid bilayer
(D) Proteins embedded in a polymer of glucose molecules

56. The suffix 'S' in ribosome unit indicates

(A) Solubility
(B) Sedimentation coefficient
(C) Surface Area
(D) Size

57. Tonoplast, also called vacuolar membrane is a differentially permeable membrane surrounds the ______.

(A) Cytoplasm (B) Vacuole
(C) Nucleus (D) Mitochondria

58. When a ripe tomato is pricked with a needle a watery fluid comes
out. This fluid is stored in

(A) Vacuole (B) Plastid
(C) Cytoplasm (D) Nucleus

59. which one of the following structures between two adjacent cells is an effective transport pathway?

(A) Plasmalemma
(B) Plasmodesmata
(C) Plastoquinones
(D) Endoplasmic reticulum

60. Cell theory was proposed by

(A) Beadle and Tatum
(B) Robert Hooke
(C) Schleiden and Schwann
(D) Leenuwenhoeck

61. Semiautonomous organelle in the cell is

(A) Peroxisomes
(B) Chloroplast
(C) Endoplasmic reticulam
(D) Golgibodies

62. The membrane around the vacuole is called

(A) Cytoplast (B) Tonoplast
(C) Amyloplast (D) Elaioplast

63. Identify the non-membraneous organelle from the following

(A) Ribosome
(B) Endoplasmic reticulam
(C) Nucleus
(D) Chloroplast

64. Microfilaments are composed mainly of a protein called

(A) Actin (B) Tubulin
(C) Myosin (D) Chitin

65. Which of the following is associated with the structure of Golgicomplex?

(A) Cristae
(B) Cisternae
(C) Annuli
(D) Quatasomes Biochemistry Quizzes

66. The subunit of prokaryotic ribosomes are

(A) 60 S+40 S (B) 70 S+30 S
(C) 60 S+30 S (D) 50 S+30 S

67. Plant cell wall mainly composed of

(A) Cellulose (B) Starch
(C) Protein (D) Lipid

68. Smooth endoplasmic reticulam is the site of

(A) protein synthesis
(B) carbohydrate synthesis
(C) amino acid synthesis
(D) Lipid synthesis

69. In higher plants, the shape of the chloroplast is

(A) Discoid (B) Cup shaped
(C) Girdle shaped (D) Ribbon shaped

70. The main function of Centrosome is

(A) Secretion
(B) Osmoregulation
(C) Protein synthesis
(D) Formation of spindle fibre

71. Assembly of two subunits 40 S and 60 S of the ribosome is

(A) 100 S unit (B) 80 S unit
(C) 70 S unit (D) 90 S unit

ANSWER SHEET

1. **A**	2. **B**	3. **C**	4. **A**	5. **A**	6. **D**	7. **B**	8. **D**	9. **B**	10. **B**
11. **C**	12. **D**	13. **A**	14. **B**	15. **A**	16. **D**	17. **B**	18. **A**	19. **B**	20. **A**
21. **A**	22. **A**	23. **B**	24. **A**	25. **B**	26. **A**	27. **C**	28. **C**	29. **B**	30. **C**
31. **B**	32. **C**	33. **A**	34. **B**	35. **C**	36. **A**	37. **D**	38. **B**	39. **D**	40. **B**
41. **A**	42. **D**	43. **D**	44. **C**	45. **B**	46. **A**	47. **C**	48. **A**	49. **A**	50. **B**
51. **D**	52. **D**	53. **B**	54. **B**	55. **C**	56. **B**	57. **B**	58. **A**	59. **B**	60. **C**
61. **B**	62. **B**	63. **A**	64. **A**	65. **A**	66. **D**	67. **A**	68. **D**	69. **A**	70. **D**
71. **B**									

Chapter 2

MOLECULAR BIOLOGY

1. The β subunit of polymerase has a function of ____________
   (A) Promoter binding
   (B) Catalytic center
   (C) Template binding
   (D) Cation binding

2. The β' subunit of polymerase has a function of ______________
   (A) Promoter binding (B) Elongation
   (C) Cation binding (D) Termination

3. The RNA polymerase core enzyme converts into a holoenzyme on the addition of the ________ subunit.
   (A) α (B) β
   (C) β' (D) σ

4. What is the function of the ω subunit of RNA polymerase?
   (A) Subunit association
   (B) Promoter binding
   (C) Initiation and elongation
   (D) Cation binding

5. Which of the following is not the part of the RNA polymerase core enzyme?
   (A) α (B) β
   (C) ω (D) σ

6. The bacterial system has ____________ RNA polymerases.
   (A) 1 (B) 2
   (C) 3 (D) 4

7. In a nucleotide, the nitrogen base is joined to the sugar molecule by
   (A) Phosphodiester bond
   (B) Glycosidic bond
   (C) Hydrogen bond
   (D) Both A & B

8. If a double stranded DNA has 20% Thymine, the percentage of Guanine in the DNA
   (A) 30% (B) 10%
   (C) 90% (D) 40%

9. If a DNA contains 1000 base pairs, what would be its length?
   (A) 3400 Å (B) 34000 Å
   (C) 6800 (D) 1000 Å

10. What is not True for DNA in prokaryotes
    (A) Present in the form of a compact structure called nucleoid
    (B) The coils are maintained by non-histone basic proteins
    (C) Found in cytoplasm in a supercoiled condition
    (D) Packaged as nucleosomes along with histones

11. Pick the right difference between a DNA and RNA
    (A) Sugar and phosphate
    (B) Sugar and purines
    (C) Purines and phosphate
    (D) Sugar and pyrimidines

12. In the following questions, a statement of Assertion (A) is followed by a statement of Reason (R).

(1) If both Assertion and Reason are true and the reason is the correct explanation of the assertion, then mark (a)

(2) If both Assertion and Reason are true but the reason is not the correct explanation of the assertion, then mark (b)

(3) If Assertion is true but Reason are false, then mark (c)

(4) If both Assertion and Reason are false, then mark (d)

Assertion: In Griffith's experiment mice were injected by a mixture of heat killed Smooth type bacteria and live Rough type bacteria. Some mice died of pneumonia and their blood contained both live Rough type bacteria and live Smooth type bacteria.

Reason: The dead Smooth type bacteria became alive and caused pneumonia. Griffith named it as transforming principle.

13. Hershey and Chase experiment proving DNA as the genetic material was based on the principle

(A) Transduction

(B) Transformation

(C) Transcription

(D) Translation

14. A bacterial colony containing DNA made up of 100% N15 nitrogen bases is allowed to replicate in a medium containing N14 bases. After one round of replication the result would be

(A) All individuals will be identical to parents

(B) All individuals will be hybrids

(C) Only 50% individuals would be hybrids

(D) All individuals would have DNA made up of 100% N14

15. Teminism is

(A) A central dogma reverse

(B) A central dogma of molecular biology

(C) A circular flow of hereditary material

(D) An effect of cytoplasm on functioning of DNA

16. Cistron is

(A) The coding sequence of DNA

(B) The functional unit of DNA molecule that codes for a particular gene product

(C) Intervening non coding sequence of DNA

(D) The sequences which are removed during RNA splicing.

17. Read the statements given below and identify the incorrect statement.

(A) The human genome contains 3164.7 million nucleotide bases.

(B) The average gene consists of 30,000 bp

(C) The total number of genes is estimated at 30,000.

(D) Chromosome Y has 231 genes

18. The coding sequences found in split genes are called

(A) Operons (B) Introns

(C) Exons (D) Cistrons

19. The removal of which enzyme affects the synthesis of hnRNA in eukaryotes

(A) RNA polymerase II

(B) RNA primase

(C) RNA polymerase III

(D) RNA polymerase I

20. Sickle cell anemia is caused

(A) When valine is replaced by glutamic acid in beta polypeptide chain

(B) When glutamic acid is replaced by valine in beta polypeptide chain

(C) When glutamic acid is replaced by valine in alpha polypeptide chain

(D) When valine is replaced by glutamic acid in alpha polypeptide chain

21. Wobble position means

(A) Base paring

(B) altered base on code

(B) third altered base on codon

(D) none of the above

22. Peptidyl transferase

(A) Is a 23s rRNA

(B) forms peptide bonds

(C) component of ribosome

(D) all the three

23. Which mRNA will be translated to a polypeptide chain containing 8 amino acids?

(A) AUGUUAAUAGACGAGUAGCGACGAUGU

(B) AUGAGACGGACUGCAUUCCCAACCUGA

(C) AUGCCCAACCGUUAUUCAUGCUAG

(D) AUGUCGACAGUCUAAAACAGCGGG

24. Arrange the following events in the order of synthesis of a protein

(i) A peptide bond forms

(ii) A tRNA matches its anticodon to the codon in the A- site

(iii) The movement of second tRNA complex from A-site to P-site

(iv) The large subunit attaches to the small subunit and the initiator tRNA fits in the P-site

(v) A small subunit binds to the mRNA

(vi) The activated amino acid tRNA complex attaches the initiation codon on mRNA

(A) iv, v, iii, ii, i, vi

(B) iv, vi, v, ii, i, iii

(C) v, iv, iii, ii, vi, i

(D) v, vi, iv, ii, i, iii

25. Select the incorrect statement out of the five given below about lac operon when Lactose is present in the medium.

(A) Gene – A gets transcribed into mRNA which produces β-galactoside permease

(B) Inducer-Repressor complex is formed

(C) Lactose inactivates repressor protein

(D) RNA polymerase transcribe Z-gene, Y-gene and A-gene

26. The percentage of human genome which encodes proteins is approximately

(A) Less than 2% (B) 5%

(C) 25% (D) 99%

27. Match the entries in column I with those of column II and choose the correct answer.

Column I	Column II
(A) Alkali treatment	(M) separation of DNA fragments on gel slab
(B) Southern blotting	(N) split DNA fragments into single strands
(C) Electrop-horesis	(O) DNA transferred to nitrocellulose sheet
(D) PCR	(P) X-ray photography
(E) Autoradio-graphy	(Q) produce fragments of different sizes
(F) DNA treated with REN	(R) DNA amplification

(A) A - N, B - Q, C - P, D - R, E- M, F - O

(B) A - P, B - R, C - M, D -O, E - N, F - Q

(C) A - Q, B - O, C - M, D - R, E - P, F - N

(D) A - N, B - O, C - M, D - R, E - P, F - Q

28. Enzyme which can break and seal the DNA strand

(A) Topoisomease II

(B) Helicase (C) Primase

(D) Restriction endonuclease

29. Which of the statements give below is correct with respect to frameshift mutation

(A) a single nucleotide base change, insertion, or deletion of the genetic material

(B) Glutamine is replaced by valine

(C) Sickle cell anemia is an example

(D) insertions or deletions of a number of nucleotides in a DNA sequence that is not divisible by three.

30. Match the names of scientists in column I with their achievements in column II and choose the correct answer given below

Column I	Column II
(A) Watson and Crick	(P) DNA fingerprinting
(B) R.W. Holley	(Q) Decipher genetic code
(C) Marshal Nirenberg	(R) Double helix of DNA
(D) Jacob and Monod	(S) Clover model of tRNA
(E) Alec Jeffrey	(T) Lac operon concept

	(A)	(B)	(C)	(D)	(E)
(A)	R	S	P	T	Q
(B)	R	S	Q	T	P
(C)	R	Q	P	T	S
(D)	R	T	S	P	Q

31. The transcription initiation factor associated with the RNA polymerase holoenzyme in prokaryotes is

(A) β (B) ω
(C) σ (D) αI

32. The stretch of codons between AUG and a stop codon is called

(A) Open reading frame
(B) TATA box
(C) Colinearity (D) Degenerate

33. The structural genes of lac operon transcribe mRNA which is

(A) Polycistronic (B) Replicative
(C) Monokaryotic (D) Monocistronic

34. If the sequence of bases in DNA is TACCGACCA, then the sequence of codons on the transcript will be

(A) ATGGCTGGT
(B) ATCCGAACU
(C) AUGGCUGGU
(D) AUGGACUAA

35. Match the following

Column I	Column II
(A) Helicase	(M) activation of amino acid
(B) Peptidyl transferase	(N) joins DNA fragments
(C) DNA polymerase	(O) unwinds DNA helix
(D) DNA ligase	(P) peptide bonds between amino acids
(E) Aminoacyl synthetase enzyme	(Q) DNA synthesis
(F) RNA primase	(R) synthesis of RNA primer

(A) A-O, B-P, C-Q, D-N, E-M, F-R
(B) A-R, B-M, C-N, D-Q, E-P, F-O
(C) A-M, B-R, C-P, D-Q, E-N, F-O
(D) A-R, B- Q, C- A, D- M, E-P, F-N

36. Genes which are active all the time synthesizing substances needed by the cell are called

(A) Cellular luxury genes
(B) Metabolic genes
(C) House keeping genes
(D) Control genes

37. The technique used to identify specific DNA sequence in bacterial colonies is

(A) Colony hybridization
(B) In situ hybridization
(C) Dot blot technique
(D) Western blotting

38. Sothern hybridization is

(A) Used to identify a specific protein
(B) Used to identify a specific DNA
(C) Used to identify a specific RNA
(D) Used to identify both DNA and RNA

39. The common feature amongst nucleus, chloroplast and mitochondria is

(A) Lamellae (B) DNA
(C) Cristae (D) All the above

40. The enzyme taking part in joining two ends of DNA is

(A) Ligase (B) Polymerase
(C) Gyrase (D) Helicase

41. The technique used to detect the presence of DNA or RNA in a non□fractionated DNA sample is

(A) Colony hybridization
(B) In situ hybridization
(C) Dot blot technique
(D) Western blotting

42. Northern hybridization is

(A) Used to identify a specific protein
(B) Used to identify a specific DNA
(C) Used to identify a specific RNA
(D) Used to identify both DNA and RNA

43. In Northern hybridization probe hybridization forms

(A) DNA : DNA hybrid
(B) RNA : DNA hybrid
(C) Both A and B
(D) None of these

44. All are differences in procedure between Northern and Southern hybridization except

(A) DBM membrane is used in northern hybridization
(B) RNA:DNA hybrids are formed in northern hybridization
(C) Initially fragments are separated by electrophoresis in northern hybridization
(D) DNA denaturation is required before blotting in Southern hybridization

45. Southern hybridization is

(A) Used to identify a specific protein
(B) Used to identify a specific DNA
(C) Used to identify a specific RNA
(D) Used to identify both DNA and RNA

46. In Western blotting

(A) Agarose gel is commonly used
(B) Polyacrylamide gel is commonly used
(C) Both A and B
(D) High resolution gels

47. For glycoproteins, most commonly used probe is

(A) Antibody (B) Lectin
(C) Antigens (D) Interferons

48. For protein detection, most commonly used probe is

(A) Antibody (B) Lectin
(C) Antigens (D) Interferons

49. Structure of DNA was given by

(A) Kornberg
(B) Nirenberg
(C) Watson and Crick
(D) Holley and Nirenberg

50. The technique used to locate specific genes in chromosomes is

(A) Colony hybridization
(B) In situ hybridization
(C) Dot blot technique
(D) Western blotting

51. The separation technique of charged molecules under the influence of electric current is called

(A) Colony hybridization
(B) Electrophoresis
(C) Dot blot technique
(D) Western blotting

52. The codon for anticodon 3' UUUA – 5' is

(A) 5' AAAU- 3' (B) 5' UUUA- 3'
(C) 3' UAAD- 5' (D) 3' AUUU- 5'

53. All viruses consists of
(A) DNA and protein coat
(B) RNA and protein coat
(C) Polysaccharides and proteins
(D) A nucleic acid and a protein coat

54. In bacteria, a small circle of DNA found outside the main chromosome is called a
(A) Plasmid (B) Protein
(C) Genetic fingerprint
(D) Episome

55. The genetic instructions for forming a polypeptide chain are carried to the ribosome by the
(A) tRNA (B) rRNA
(C) mRNA (D) DNA

56. The DNA of an elephant and the DNA of a mango tree will probably differ in all of the following
(A) Sequence of DNA nucleotides
(B) Length of DNA molecules
(C) Number of DNA molecules
(D) Kinds of nucleotides utilized in forming DNA

57. The *lac* operon in *E. coli* is involved in
(A) Regulating the expression of a gene
(B) Controlling DNA replication
(C) Regulating the translation of mRNA
(D) Controlling the formation of ribosomes

58. Genetic variation can be introduced into bacteria by all of the following methods except:
(A) Transduction (B) Mutation
(C) Transformation
(D) DNA amplification

59. Similarity between DNA and RNA is that both have
(A) Similar sugars
(B) Similar mode of replication
(C) Similar pyrimidines
(D) Polymers of nucleotides

60. All of the following enzymes are involved in DNA replication except:
(A) Helicase (B) Primase
(C) DNA polymerase
(D) RNA polymerase

61. The term cistorn, muton and recon were introduced by
(A) Watson and Crick
(B) S. Benzer
(C) Meselson (D) Morgan

62. Extranuclear genetic material is found in
(A) Plastid and nucleus
(B) Mitochondria and plastids
(C) Nucleus and cytoplasm
(D) Mitochondria and nucleus

63. The molecular formulae of deoxyribose sugar and ribose sugar respectively are
(A) C H O and CH O
(B) C H O and CH O
(C) C H O and CH O
(D) C H O and CH O

64. The nitrogen bases which pair with two hydrogen bonds are
(A) Adenine and thymine
(B) Adenine and Cytosine
(C) Cytosine and guanine
(D) Cytosine and adenine

65. The distance between two successive nitrogenous base pairs is
(A) 34 Å (B) 36 Å
(C) 20 Å (D) 3.4 Å

66. In nucleoside, nitrogen base is attached to pentose sugar at
(A) Carbon – 1 of pentose sugar
(B) Carbon – 2 of pentose sugar
(C) Carbon – 4 of pentose sugar
(D) Carbon – 5 of pentose sugar

67. If the strand of DNA has 35 nucleotide how many phosphodiester bonds would exist

(A) 34 (B) 35

(C) 24 (D) 70

68. In eukaryotic DNA replication, lagging strand is formed by

(A) RNA fragments

(B) Okazaki fragments

(C) DNA fragments

(D) Nucleotide fragments

69. The enzyme DNA polymerase can work only in

(A) 3' → 5' direction

(B) 5' → 3' direction

(C) Both the direction

(D) 5' → 5' direction

70. Enzyme required for removing RNA primer during DNA replication is

(A) DNA primase

(B) DNA ligase

(C) DNA polymerase I

(D) DNA polymerase III

71. During DNA replication, the reunion or recoiling of separated DNA strand is prevented by

(A) Helix destabilizing protein

(B) Single strnad binding protein

(C) Rep protein

(D) Both (A) and (B)

72. The enzyme that cuts the bonds of DNA molecule at the origin of replication is

(A) Endonuclease

(B) DNA polymerase

(C) DNA gyrase

(D) DNA ligase

73. The cellular composition of m-RNA is

(A) 5-10% (B) 3-5%

(C) 10-20% (D) 70-80%

74. Formation of mRNA from DNA is called

(A) Transformation

(B) Transduction

(C) Traslation

(D) Transcription

75. The ratio of purines and pyrimidines in mRNA is not 1:1 because the nitrogenous bases are

(A) Unpaired

(B) Paired

(C) Paired only in loops

(D) Paired in stems

76. The codons which may present at 3¢ end of mRNA

(A) UAA (B) UAG

(C) UGA (D) Any one of these

77. A nucleotide is formed of

(A) Purine, Pyrimidine and phosphate

(B) Purine, Sugar and phosphate

(C) Nitrogen base, Sugar and phosphate

(D) Pyrimidine, Sugar and phosphate

78. A riboside is

(A) Base + phosphate

(B) Ribose + phosphate

(C) Ribose + phosphate + base

(D) Ribose + base

79. A segment of DNA has 120 adenine and 120 cytosine bases. The total number of nucleotides present in the segment is

(A) 120 (B) 240

(C) 60 (D) 480

80. Adenine is

(A) Purine (B) Pyrimidine

(C) Nucleoside (D) Nucleotide

81. Amino acid binding site of tRNA is

(A) 5'end (B) Anticodon loop

(C) DHU loop (D) -CCA 3'end

82. Anticodon occurs in

(A) tRNA (B) mRNA
(C) mtRNA (D) rRNA

83. Bacterial plasmid contains

(A) RNA
(B) RNA + protein
(C) DNA
(D) Photosynthetic structures

84. Base pairs present in one turn of DNA are

(A) 12 (B) 11
(C) 10 (D) 9

85. Distance between two base pairs of DNA is

(A) 34 nm (B) 3.4 nm
(C) 0.68 nm (D) 0.34 nm

86. Distance between two strands of DNA is

(A) 34 Å (B) 20 Å
(C) 3.4 Å (D) 340 Å

87. DNA and RNA are similar in having

(A) Similar nucleotides
(B) Similar pyrimidines
(C) Similar purines
(D) Similar sugars

88. DNA and RNA show similarity in having

(A) Polymers of nucleotides
(B) Similar pyrimidines
(C) Double strands
(D) Similar sugars

89. DNA does not occur in

(A) Nucleus (B) Ribosomes
(C) Mitochondria (D) Plastids

90. DNA duplex shows

(A) Right handed coiling and parallel
(B) Right handed coiling and antiparallel
(C) Left handed coiling and antiparallel
(D) Left handed coiling and parallel

91. DNA replication in eucaryotes commences

(A) From both ends of a chromosome simultaneously
(B) Several sites along DNA of a chromosome simultaneously
(C) From centromere to either end
(D) From one end of chromosome to the other

92. DNA replication is

(A) Conservative and discontinuous
(B) Semiconservative and semidiscontinuous
(C) Semiconservative and discontinuous
(D) Conservative

93. DNA resembles RNA as both have

(A) Polymers of nucleotides
(B) Similar sugars
(C) Similar pyrimidine bases
(D) Ability to replicate

94. DNA sequence is ATG. What would be the sequence of bases in anticodon of tRNA

(A) ATG (B) AUG
(C) UAC (D) TAC

95. DNA sequence is TAG. What shall be the sequence in anticodon of tRNA

(A) UAG (B) ATC
(C) ATG (D) UAC

96. DNA sequence of ATTCGATG is transcribed as

(A) AUUCGAUG (B) UAAGCUAC
(C) CAUCGAAU (D) GUAGCUUA

97. DNA strand with nitrogen base sequence ATTGCC will have sequence in mRNA

(A) UAACGC (B) ATCGCC
(C) ATTGCA (D) AGGACC

98. Double chained DNA strand is made radioactive in both its chains. It is allowed to replicate twice in non-radioactive medium. The result would be

(A) All strands have radioactivity
(B) Half the strands have radioactivity
(C) Three strands have radioactivity
(D) Radioactivity is absent in all strands

99. Functional unit of gene that specifies synthesis of one polypeptide is

(A) Codon (B) Cistron
(C) Recon (D) Muton

100. Gene is segment of

(A) RNA (B) DNA
(C) RNA or DNA
(D) Both DNA and RNA

101. Genetically engineered bacteria are being used in commercial production of

(A) Melatonin (B) Testoteron
(C) Human insuline (D) Thyroxine

102. Haploid plant cultures are got from

(A) Leaves (B) Root tip
(C) Pollen grain (D) Buds

103. Hydrogen bonds present between cytosine and guanosine are

(A) 2 (B) 3
(C) 1 (D) 4

104. Protein helping in opening of DNA double helix in front of replication fork is

(A) DNA gyrase
(B) DNA Polymerase I
(C) DNA ligase
(D) DNA topoisomerase

105. Pyrimidine base present in RNA in place of thymine of DNA is

(A) Uracil (B) Adenine
(C) Cytosine (D) Guanine

106. RNA that picks up specific amino acid from amino acid pool of cytoplasm to carry it to ribosome during protein synthesis is

(A) tRNA (B) mRNA
(C) rRNA (D) gRNA

107. Semiconservative DNA/Chromosome replication using 14N was demonstrated by

(A) Messelson (B) Tylor
(C) Messelson and stahl
(D) Hershey and Chase

108. The term "Biotechnology" was first coined by ....

(A) Neuberg
(B) Lipmann
(C) Frederick Sanger's
(D) Karl Ereky

109. Discovery of the double-helix structure of DNA

(A) 1940 (B) 1980
(C) 1990 (D) 1953

110. Chromatin assemblage contain

(A) RNA + DNA
(B) Lipids + DNA
(C) Carbohydrates + RNA
(D) Protein + DNA

111. Choose the width of a DNA molecule

(A) 5 nm (B) 4 nm
(C) 2 nm (D) 0.1 nm

112. A segment of DNA that is transcribed to produce one RNA molecule is called

(A) Codon
(B) Translation unit
(C) Gene expression
(D) Transcription Unit

113. In DNA the nitrogenous base, Adenine always base pairs with

(A) Thymine (B) Guanine
(C) Cytocin (D) Uracil

114. Prokaryotic genome is a

(A) Haploid (B) Diploid
(C) Polyploid (D) Multiploid

115. The Sulphur containing amino acid is

(A) Methionine (B) Lysine
(C) Histidine (D) arginine

116. The amino acid Methionine is encoded by which codon

(A) UGG (B) AUG
(C) CAU (D) GGG

117. In eukaryotes Transcription and RNA processing are completed in
(A) Nucleus
(B) Cytoplasm
(C) Endoplasmic reticulum
(D) Golgi body

118. In eukaryotes the functional half life of mRNA is
(A) 6 hours (B) 8 hours
(C) 12 hours (D) 24 hours

119. A segment of DNA that is transcribed to produce one RNA molecule is called
(A) Codon (B) Translation unit
(C) Gene expression
(D) Transcription Unit

120. which type of DNA structure found in most of the living organisms ___________
(A) Z- Type (B) S-Type
(C) B –type DNA (D) ssDNA

121. Activation of amino acid is carried out by .... enzyme.
(A) DNA Polymerase
(B) Amino acyle tRNA synthatase
(C) peptidyl transferase
(D) RNA polymerase

122. .... enzyme is involved in transcription.
(A) DNA Polymerase
(B) RNA Polymerase
(C) Amino acyl tRNA transferase
(D) Restriction enzymes

123. ....is an inducer of the lac operon.
(A) Lactose (B) cAMP
(C) ATP (D) Allolactose

124. .... enzyme relaxes the supercolling of double stranded DNA molecule
(A) DNA Polymerase
(B) Topoisoemerase
(C) DNA helicase
(D) DNA Ligase

125. The nuclear DNA-protein complex is called ....
(A) Chromosome (B) Nucleoprotein
(C) Chromatin (D) Nucleus

126. Each amino acid in a protein is specified by........................
(A) several genes.
(B) a promoter.
(C) an mRNA molecule.
(D) a codon.

127. .... one of the following is not a type of RNA?
(A) . nRNA (nuclear RNA)
(B) . mRNA (messenger RNA)
(C) . rRNA (ribosomal RNA)
(D) . tRNA (transfer RNA)

128. DNA is made up of building blocks called ....
(A) . proteins. (B) . bases.
(C) . nucleotides. (D) . deoxyribose.

129. Euchromatin is
(A) Lightly packed form of chromatin
(B) densely packed form of chromatin
(C) Concentrated form of chromatin
(D) elongated form of chromatin

130. The usual form of DNA noticed in most of the organisms is
(A) A form (B) B form
(C) Z form (D) Linear form

131. Endonuclease enzyme hydrolyzes
(A) Nucleotide from 3' end
(B) Nucleotide from 5' end
(C) Nucleotides from 5' and 3' ends
(D) Nuclotide from the interio

132. Which of the following nucleic acids has a left-handed helix
(A) Z – DNA (B) A – DNA
(C) B – DNA (D) RNA

133. The term Nucleic acid was introduced by

(A) Meisher (B) Robert Brown
(C) Lipmann (D) Altmann

134. The enzyme used for joining two ends of DNA is

(A) DNA polymerase
(B) DNA ligase
(C) Primase
(D) Topoisomeras

135. The segments synthesized in lagging strand during replication of DNA is

(A) Satellite segments
(B) Double helix segments
(C) Okazaki segments
(D) Kornbeg segments

136. The simplest amino acid is

(A) Proline (B) Methionine
(C) Glycine (D) Serin

137. Sulphur containing amino acids are

(A) Cysteine and Methionine
(B) Methionine and threonine
(C) Cysteine and threonine
(D) Cysteine and serine

138. During translation of which of the following event will occur

(A) Ribosomes will read the information on DNA
(B) Ribosomes will read the information of mRNA
(C) Ribosomes will read the information of rRNA
(D) Ribosomes will read the information of tRNA

139. The type of sugar in DNA is

(A) Triose (B) Tetrose
(C) Pentose (D) Hexos

140. The two adjacent strands of DNA can be joined by

(A) Covalent bond
(B) Hydrogen bond
(C) Wanderval force
(D) Phosphodiester bond

141. Okazaki fragments are......................

(A) Synthesized in the 3' to 5' direction.
(B) Found on the lagging strand.
(C) Found on the leading strand.
(D) Assembled as continuous replication.

142. Beadle and Tatum's experiment showed that each enzyme is specified by a single ....

(A) Chromosome (B) Gene
(C) Nucleotide (D) Mutation

143. The average pH of the double distilled water is...........

(A) 6 (B) 8
(C) 5 (D) 7

144. Plants having nucleus of one species but cytoplasm from both parental species called as................

(A) Cybrids (B) Hybrids
(C) Vybrids (D) Sybrids

145. Match the following

(a) Universal solvent	(i) Virus
(b) Protoplast culture	(ii) Sucrose
(c) Carbon source	(iii) Somatic hybrid
(d) Transduction	(iv) Water

(A) i, ii, iii, iv (B) iii, ii, iv,i
(C) iii, iv, i, ii (D) iv, iii, ii, i

146. Match the following

(a) Transformation	(i) Seed Dormancy
(b) cDNA	(ii) Pneumococcal
(c) Transformation factor	(iii) Reverse transcriptase
(d) Embryo culture	(iv) Bacteria

(A) i, ii, iii, iv (B) iii, ii, iv, i
(C) iii, iv, i, ii (D) iv, iii, ii, i

147. Match the following

(A) In direct organogenesis	(i) UAG, UAA, and UGA
(B) ORF	(ii) Essential genes
(C) Genomic DNA	(iii) Coding region
(D) Stop codon	(iv) Somatic embryo

(A) i, ii, iii, iv (B) iii, ii, iv, i
(C) iii, iv, i, ii (D) iv, iii, ii, i

148. Match the following

(A) Sulphur contain amino acid	(i) Chloroplast DNA
(B) Disinfection	(ii) All living organism
(C) B-type DNA	(iii) Sodium hypochlorite
(D) Organelle DNA	(iv) Cysteine and methionine

(A) iv, iii, ii, i (B) iii, ii, iv, i
(C) iii, iv, i, ii (D) i, ii, iii, iv

149. Match the following

(A) Agar-Agar	(i) Cotton
(B) Topoisomerase	(ii) DNA content
(C) C-value	(iii) Unwinding DNA
(D) Polyploidy	(iv) Gelling agent

(A) iv, iii, ii, i (B) iv, iii, ii, i
(C) i, ii, iii, iv (D) iii, iv, i, ii

150. The Structure of tRNA is ________.

(A) Single stranded
(B) Clover leaf
(C) Double stranded
(D) Linear

151. The semiconservative replication of DNA was first propsed by ________.

(A) Watson and Crick
(B) Wilson and Franklin
(C) Meselson and Stahl
(D) Stanley and Cohen

152. During DNA replication process the enzyme involved in unwinding of DNA strands________.

(A) Helicase (B) Topoisomerase
(C) Primase (D) DNA ligase

153. ________ is responsible for the accreditation of test laboratories.

(A) NCS-TCP Management cell (NMC)
(B) Accreditation panel
(C) Appellate authority (AA)
(D) Referral laboratories

154. C- value related to _____

(A) DNA size
(B) Chromosome number
(C) DNA content
(D) Genome size

155. Give an example of stress amino acid in plants _______

(A) Cysteine and methionine
(B) Methionine
(C) Proline
(D) UAA, UGA, UAG

156. Give an example of sulphur contain amino acid _______

(A) Cysteine and methionine
(B) Methionine
(C) Proline
(D) UAA, UGA, UAG

157. The activity of RNA polymerase was first discovered by

(A) Spiegelman (B) Wess
(C) Khorana (D) Sanger

158. During Transcription only one strand of a DNA molecule is transcribed. This strand is called

(A) Antisense strand
(B) Coding strand
(C) Sense strand
(D) Non coding strand

159. The location of the RNA polymerase I enzyme is

(A) Nucleolus (B) Nucleoplasm
(C) Cytoplasm (D) Vacoule

160. The mRNA carries genetic information for the protein to be synthesized in the cell in the form of ------------

(A) Codons (B) Triplet code
(C) Translation
(D) Reverse Transcription

161. In prokaryotes the initiation factor IF3 binds to the sub unit of

(A) 30 S (B) 50 S
(C) 40 S (D) 60 S

162. Which one the following is present only in eukaryotic mRNA

(A) 5' CAP
(B) Untranslated 5' leader sequence
(C) Coding region
(D) Poly A tail

163. In eukaryotes the function of elongation factor elF3 is

(A) Binding mRNA, maintains free 40 S subunits
(B) Binding mRNA at 5' CAP and unwinding
(C) Assists mRNA unwinding
(D) Binding Met-tRNAi

164. What is the relative size of prokaryotic genome?

(A) 600kb to 9.5 Mb
(B) 3 Mb to 140000Mb
(C) 600kb to 300 Mb
(D) 30 Mb to 140000Mb

165. What is the relative size of eukaryotic genome?

(A) 600kb to 9.5 Mb
(B) 3 Mb to 140000Mb
(C) 600kb to 300 Mb
(D) 30 Mb to 140000Mb

166. What is the average gene size/number of prokaryotic genome?

(A) 950 bp / 4300
(B) 2500 bp / 19000
(C) 600 bp / 4300 Mb
(D) 250 bp / 140000

ANSWER SHEET

1. **B**	2. **A**	3. **B**	4. **A**	5. **D**	6. **C**	7. **B**	8. **A**	9. **A**	10. **D**
11. **D**	12. **C**	13. **A**	14. **B**	15. **A**	16. **B**	17. **B**	18. **C**	19. **A**	20. **B**
21. **B**	22. **D**	23. **B**	24. **D**	25. **A**	26. **A**	27. **D**	28. **A**	29. **D**	30. **B**
31. **C**	32. **A**	33. **A**	34. **C**	35. **A**	36. **C**	37. **A**	38. **B**	39. **B**	40. **A**
41. **C**	42. **C**	43. **C**	44. **C**	45. **A**	46. **B**	47. **B**	48. **A**	49. **C**	50. **B**
51. **B**	52. **A**	53. **D**	54. **A**	55. **C**	56. **D**	57. **A**	58. **D**	59. **D**	60. **D**
61. **B**	62. **B**	63. **B**	64. **A**	65. **D**	66. **A**	67. **A**	68. **B**	69. **B**	70. **C**
71. **D**	72. **A**	73. **B**	74. **D**	75. **A**	76. **D**	77. **C**	78. **D**	79. **D**	80. **A**
81. **D**	82. **A**	83. **C**	84. **C**	85. **D**	86. **B**	87. **C**	88. **A**	89. **B**	90. **B**
91. **B**	92. **C**	93. **A**	94. **B**	95. **A**	96. **B**	97. **A**	98. **B**	99. **B**	100. **B**
101. **C**	102. **C**	103. **B**	104. **A**	105. **A**	106. **A**	107. **C**	108. **D**	109. **D**	110. **D**

111. **C**	112. **A**	113. **A**	114. **A**	115. **A**	116. **B**	117. **A**	118. **A**	119. **D**	120. **C**
121. **B**	122. **B**	123. **A**	124. **C**	125. **C**	126. **D**	127. **A**	128. **C**	129. **A**	130. **B**
131. **D**	132. **A**	133. **D**	134. **B**	135. **C**	136. **C**	137. **A**	138. **B**	139. **C**	140. **B**
141. **B**	142. **B**	143. **D**	144. **A**	145. **D**	146. **D**	147. **C**	148. **C**	149. **D**	150. **B**
151. **C**	152. **A**	153. **A**	154. **C**	155. **C**	156. **A**	157. **B**	158. **C**	159. **A**	160. **B**
161. **A**	162. **D**	163. **D**	164. **A**	165. **B**	166. **A**				

Chapter 3

GENETIC ENGINEERING

1. The first crop plant genome sequenced
   (A) Maize (B) Wheat
   (C) Rice (D) Barley
2. Urea and Formamide can:
   (A) Increase Tm of DNA
   (B) Decrease the Tm of DNA
   (C) Helps faster re-annealing of DNA
   (D) Does not effects denaturation of DNA
3. The chromogenic substrate of X-gal is:
   (A) 5-chloro-4-bromo-3indolyl-beta-D-galactoside
   (B) 5-bromo-4-chloro-3indolyl-beta-D-galactoside
   (C) 5-chloro-5-bromo-3indolyl-beta-D-galactoside
   (D) 5-bromo-5-chloro-3indolyl-beta-D-galactoside
4. The first transgenic plant to be produced:
   (A) Rice (B) Maize
   (C) Cotton (D) Tobacco
5. Which of the following molecules is useful as a measure of relatedness beyond species level between organisms:
   (A) mRNA (B) tRNA
   (C) rRNA (D) miRNAPaper
6. Explants used for tissue culture can be taken from:
   (A) Bud
   (B) Root tip
   (C) Nodal stem segments
   (D) Entire shoot
7. Indole -3-acetic acid is:
   (A) A gibberellin
   (B) A cytokinin
   (C) An auxin
   (D) An abscisic acid
8. A low auxin:cytokinin ratio leads to:
   (A) Shoot formation
   (B) Root formation
   (C) Fruit formation
   (D) Flower formation
9. Hairy root cultures are produced by using:
   (A) *Agrobacterium tumefaciens*
   (B) *Agrobacterium rhizogenes*
   (C) *Agrobacterium vitis*
   (D) *Agrobacterium rubi*
10. Klenow fragment is derived from:
    (A) DNA Pol-I
    (B) RNA Pol II
    (C) DNA Ligase
    (D) Reverse Transcriptase
11. Polyhedrin protein is associated with one of the following viruses:
    (A) HIV
    (B) HSV
    (C) RSV
    (D) NPV

12. The year of launching of Human Genome Project and 1completion of the rough draft of the sequence was in:
(A) 1992 and 2000
(B) 1990 and 2000
(C) 1989 and 2001
(D) 1992 and 2002

13. Which of the following is used as a fusogen:
(A) Polyethylene glycol
(B) CaMV
(C) Sendai Virus
(D) Adenovirus

14. Restriction endonucleases are usually supplied in:
(A) 10% glycerol
(B) 30% glycerol
(C) 50% glycerol
(D) 80% glycerol

15. The main difference between a plasmid and an episome is:
(A) Plasmid is circular episome is linear
(B) Episome is circular, plasmid is linear
(C) Plasmid can be insert into host genome, episomes are not
(D) Episomes can be inserted into the host genome

16. The virus mediated gene transfer using genetically modified bacteriophages is called
(A) Transfection
(B) Transduction
(C) Transformation
(D) Conjugation

17. The ability of cells to take up DNA fragments from surrounding is called
(A) Transfection
(B) Transduction
(C) Transformation
(D) Conjugation

18. Which of the following bacterium is considered as 'natural genetic engineer'
(A) *Agrobacterium tumefaciens*
(B) *Agrobacterium radiobactor*
(C) *Psueudomonas putida*
(D) *Thermus aquaticus*

19. The removal or replacement of tumor causing genes from Ti plasmid is termed as
(A) gne replacement
(B) Diarming
(C) Inertional inactivation
(D) gne displacement

20. *Agrobacterium tumefaciens* is a
(A) Gram negative soil bacterium causing crown gall disease in dicots
(B) Gram negative soil bacterium causing crown gall disease in monocots
(C) Gram positive soil bacterium causing crown gall disease in dicots
(D) Gram positive soil bacterium causing crown gall disease in dicots

21. Which of the following statements are true for agrobacterium mediated gene transfer
(A) Vir genes are essential for gene transfer
(B) T-DNA borders are essential for gene transfer
(C) Both A and B
(D) Ori is essential for gene transfer

22. Ti plasmid vectors include
(A) Binary vectors and cointegrate vectors
(B) Cointegrate vectors and multiple vectors
(C) Multiple vectors and binary vectors
(D) Ti plasmid based vector

23. Which of the following chemical enhances *vir* gene expression
(A) Cyanidin (B) Glutennin
(C) Acetosyringone
(D) Dextran

24. Chemicals used for gene transfer methods include

(A) Poly ethylene glycol

(B) CaCl2

(C) Dextran

(D) All of the above

25. Introduction of DNA into cells by exposing to high voltage electric pulse is

(A) Electrofusion (B) Elctrofision

(C) Elctrolysis (D) Electroporation

26. The transformation method that uses tungsten or gold particle coated with DNA accelerated at high velocity is called

(A) Acceleration method

(B) High velocity method

(C) Particle gun delivery method

(D) DNA particle delivery method

27. The method widely used for transforming *invitro* animal cell cultures that uses lipid vescicles or liposomes

(A) Lipotransformation

(B) Liposome mediated transformation

(C) Lipofection

(D) Lipid mediated DNA transfer

28. DNA solution injected directly into the cell using micromanipulators is called

(A) Macroinjection

(B) Micromanipulator mediated DNA delivery

(C) Microfection

(D) Microinjection

29. Fibre mediated DNA delivery uses

(A) Silicon carbide fibres that will create pores in the membrane

(B) Aluminium carbide fibres that will create pores in the membrane

(C) Boron carbide fibres that will create pores in the membrane

(D) Lead carbide fibres that will create pores in the membrane

30. The injection of DNA into developing inflorescence using a hypodermic syringe is called

(A) Macroinjection

(B) Micromanipulator mediated DNA delivery

(C) Microfection

(D) Microinjection

31. Any DNA molecule that has the ability to replicate in an appropriate host cell, to which the desired gene are integrated for cloning, is called as

(A) Plasmid (B) Linker

(C) Vector (D) Adapter

32. Which of the following statement is true

(A) A vector may undergo self ligation

(B) A vector may have selectable markers

(C) A vector should have a promoter

(D) A vector should have unique restriction sites

33. Expression vectors differ from a cloning vector in having

(A) An origin of replication

(B) Suitable marker genes

(C) Unique restriction sites

(D) Control elements

34. Extra chromosomal, doublé stranded, circular DNA molecule present in bacteria which is widely used as vector is called

(A) Phagemid (B) Cosmid

(C) Plasmid (D) Bacterial vectors

35. The first engineered plasmid vector is

(A) pBR 322 (B) pUC vectors

(C) pSC101 (D) pUC 19

36. The most popular and widely used engineered plasmid vector is

(A) pBR 322 (B) pUC vectors

(C) pSC101 (D) pUC 19

37. In pBR 322, pBR stands for
(A) Plasmid bacterial recombination
(B) Plasmid bacterial replication
(C) Plasmid Boliver and Rodriguez
(D) Plasmid Baltimore and Rodriguez

38. Phage M13 vectors are widely used for
(A) Obtaining single stranded copies of cloned DNA suitable for DNA sequencing
(B) Obtaining double stranded copies of cloned DNA suitable for DNA sequencing
(C) Obtaining fragments of cloned DNA suitable for DNA sequencing
(D) Obtaining double stranded copies of cloned DNA suitable for electrophoresis

39. Cosmid is a plasmid with
(A) A minimum of 250 bp of lambda DNA that includes *cos* site
(B) A minimum of 250 bp of M13 DNA that includes *cos* site
(C) A minimum of 100 bp of lambda DNA that includes *cos* site
(D) A minimum of 250 bp of T4 phage DNA that includes *cos* site

40. Vectors designed to replicate in cells of two different species are called
(A) Phasmids
(B) Transfer vectors
(C) Shuttle vectors
(D) Phagemids

41. Autonomously replicating sequences (ARS) is a characteristic feature of
(A) Plasmid vectors
(B) Phage vectors
(C) *E. coli* vectors
(D) Yeast vectors

42. Which of the following are vectors for animals
(A) SV 40 vectors and Bovine papillomavirus vectors
(B) CMV vectors and Gemini vectors
(C) Lambda phage and M13 phage vectors
(D) Retro vectors and Adeno vectors

43. The vectors commonly used for sequencing human genome
(A) Yeast artificial chromosome (YAC)
(B) Plasmid
(C) CMV vectors
(D) M13 vectors

44. Ti plasmids that is used as a plant vector is obtained from
(A) *Agrobacterium tumefaciens*
(B) *Agrobacterium rhizhogenes*
(C) *Agrobacterium radiobactor*
(D) *Thermus aquaticus*

45. The most common vectors for plants are
(A) SV 40 vectors and Bovine papillomavirus vectors
(B) Cauliflower mosaic virus (CMV vectors) and Gemini vectors
(C) Lambda phage and M13 phage vectors
(D) T4 phage vectors

46. Cutting certain genes out of molecules of DNA requires the use of special
(A) Degrading nucleases
(B) Restriction endonucleases
(C) Eukaryotic enzymes
(D) Viral enzymes

47. Which of the following cannot be used as a vector?
(A) Phage (B) Plasmid
(C) Bacterium (D) Cosmid

48. Transposon is known as
(A) IS element
(B) Jumping gene
(C) Conservative gene
(D) Co integrate gene

49. A DNA library is

(A) A general collection of all genes sequenced thus far

(B) A collection of DNA fragments that make up the entire genome of a particular organism

(C) A DNA fragment inserted into a vector

(D) All DNA fragments identified with a probe

50. A probe is used in which stage of genetic engineering?

(A) Cleaving DNA

(B) Recombining DNA

(C) Cloning

(D) Screening

51. The enzyme used in the polymerase chain reaction is

(A) Restriction endonuclease

(B) Reverse transcriptase

(C) DNA polymerase

(D) RNA polymerase

52. Inserting a gene encoding a pathogenic microbe's surface protein into a harmless virus produces a

(A) Piggyback vaccine

(B) Virulent virus

(C) Active disease-causing pathogen

(D) Pharmaceutical human protein

53. Although the Ti plasmid has revolutionized plant genetic engineering, one limitation of its use is that it

(A) Cannot infect broadleaf plants

(B) Cannot be used on fruit-bearing plants

(C) Cannot transmit prokaryotic genes

(D) Does not infect cereal plants such as corn and rice

54. Which of the following is *not* an application of genetic engineering in plants?

(A) Nitrogen fixation

(B) DNA vaccines

(C) Resistance to glyphosate

(D) Production of insecticidal proteins in plants

55. The gene formed by the joining of DNA segments from two different sources are called as

(A) Recombinant gene

(B) Joined gene

(C) Both A and B

(D) Chimaeric gene

56. Which of the following enzyme is used to cut DNA molecule in rDNA technology

(A) Ligase

(B) Phosphatase

(C) Ribonuclease

(D) Restriction enzymes

57. The most important discovery that lead to the development of rDNA technology was

(A) Double helix model of Watson and Crick

(B) Discovery of restriction enzymes

(C) Discovery of ligase enzyme

(D) Discovery of plasmids

58. Who discovered restriction enzymes

(A) Nathan, Arber and Smith in 1970

(B) Watson, Crick and Wilkins in 1970

(C) Boyer and Cohen in 1975

(D) Paul Berg in 1975

59. Who created the first rDNA molecule

(A) Nathan, Arber and Smith

(B) Watson, Crick and Wilkins

(C) Boyer and Cohen

(D) Paul Berg

60. The DNA molecule to which the gene of insert is integrated for cloning is called

(A) Carrier

(B) Transformer

(C) Vector

(D) Plasmid

61. The DNA segment to be cloned is called

(A) Gene segment
(B) DNA fragment
(C) DNA insert
(D) Gene of interest

62. Two bacteria most useful in genetic engineering are

(A) *Rhizobium* and *Azobacter*
(B) *Nitrosomonas* and *Klebsilla*
(C) *Escherichia* and *Agrobacterium*
(D) *Rhizobium* and *Diplococcus*

63. The first successful transformation of rDNA molecule into a bacterium was carried out by

(A) Nathan, Arber and Smith
(B) Watson, Crick and Wilkins
(C) Boyer and Cohen
(D) Paul Berg

64. The plasmid used by Cohen and Boyer for their transformation experiment was

(A) pSC 101 (B) PUC 17
(C) pBR 322 (D) *E. coli* plasmids

65. The mechanism of intake of DNA fragments from the surrounding medium by a cell is called

(A) Transformation
(B) Transduction
(C) Both A and B
(D) Conjugation

66. Which enzyme is useful in genetic engineering?

(A) DNAase
(B) Amylase
(C) Lipase
(D) Restriction endonuclease

67. Paul Berg's gene splicing experiment created thefirst rDNA molecule which was a

(A) A T4 phage fragment incorporated into SV40 vector
(B) A lambda phage fragment incorporated into SV40 vector
(C) A T4 phage fragment incorporated into pSC 101 vector
(D) A lambda phage fragment incorporated into pSC 101 vector

68. Which of the following enzyme is used to cut DNA molecule internally

(A) Restriction enzymes
(B) Restriction endonuclease
(C) Restriction exonuclease
(D) Ribonuclease H

69. Which type of restriction enzymes are commonly used in rDNA technology

(A) Type I (B) Type II
(C) Type III (D) Type IV

70. Which of the following enzyme is used to join two DNA molecule

(A) Nuclease
(B) Restriction enzymes
(C) Lyases
(D) Ligases

71. Which is the enzyme used to remove phosphate group from the 5' end of the DNA

(A) Restriction enzymes
(B) Alkaline phosphatase
(C) Polynucleotide kinase
(D) Ribonuclease HPractice Tests

72. Transgenic plants are developed by

(A) Introducing foreign genes
(B) Introducing gene mutations
(C) Deleting certain chromosomes parts
(D) Stopping spindle formation

73. Cry genes or Bt genes are obtained from

(A) Cotton pest
(B) Tobacco plant
(C) *Bacillus thuringiensis*
(D) *E. coli*

74. Klenow enzyme is the product of enzymatic breakdown of

(A) DNA polymerase I
(B) DNA polymerase II
(C) DNA polymerase III
(D) RNA polymerase

75. Selective degradation of single stranded DNA is carried out by the enzyme

(A) Nuclease (B) Ribonuclease
(C) SI nuclease
(D) Deoxy ribonuclease

76. Which of the following is an RNA dependent DNA synthetase

(A) DNA polymerase I
(B) DNA polymerase II
(C) Reverse transcriptase
(D) All of these

77. Which of the following is a thermo stable DNA polymerase

(A) Taq polymerase
(B) Vent polymerase
(C) pfu polymerase
(D) DNA polymerase I

78. Western blotting is the technique for the detection of

(A) Specific DNA in a sample
(B) Specific RNA in a sample
(C) Specific protein in a sample
(D) Specific glycolipid in a sampleBe the first of your friends to like this

79. Arrange the following in correct order

1. Southern Blotting ----A. Alwine
2. Western blotting ----B. E.M.Southern
3. Northern Blotting----C. A. Jeffrey
4. DNA fingerprinting---- D. Towbin

(A) 1-A, 2-C, 3-D, 4-B
(B) 1-B, 2-D, 3-A, 4-C
(C) 1-B, 2-A, 3-D, 4-C
(D) 1-B, 2-C, 3-A, 4-C

80. Labelled Antibodies are used to detect

(A) Detect the presence of a particular DNA molecule in Southern blotting
(B) Detect the presence of a particular RNA molecule in Southern blotting
(C) Detect the presence of a particular protein molecule in Southern blotting
(D) Detect the presence of a particular protein molecule in Western blotting

81. Which of the following technique is suitable for identifying mRNA molecule in a sample

(A) Western blotting
(B) Southern blotting
(C) Northern blotting
(D) Eastern blotting Multiple Choice

82. Aminobenzyloxymethyl filter paper is commonly used for transfer in

(A) Western blotting
(B) Southern blotting
(C) Northern blotting
(D) Dot blotting

83. Arrange the following in correct order

1. Southern Blotting	(A) RNA-DNA hybrid
2. Western blotting	(B) DNA-DNA hybrid
3. Northern Blotting	(C) Southern blotting
4. DNA fingerprinting	(D) Antigen antibody reaction

(A) 1-A, 2-C, 3-D, 4-B
(B) 1-B, 2-D, 3-A, 4-C
(C) 1-B, 2-A, 3-D, 4-C
(D) 1-B, 2-C, 3-A, 4-C

84. A vector should have which of the following properties

(i) MCS; (ii) Small size, (iii) Multiple ori (iv) low replication speed

(A) i, ii, iii (B) ii, iii, iv
(C) i, ii, iv (D) i, iii, iv

85. DNA staining is done by
(A) Cystal violet
(B) Giemsa staining
(C) Methylene blue
(D) Feulgen staining

86. Which of the following is not a restriction endonuclease?
(A) Eco RI (B) DNA ligase
(C) Hind III (D) Bam H1

87. Use of alkaline phosphatase is to
(A) Add phosphate molecules to 3'end
(B) Add phosphate molecules to 5'end
(C) Remove terminal phosphate from both end
(D) Remove terminal phosphates from 5'end

88. The uptake of plasmid DNA into bacterial cell is facilitated in the presence of
(A) Calcium chloride
(B) Magnesium chloride
(C) Potassium chloride
(D) Sodium chloride

89. Taq polymerase is used in PCR because of its
(A) Low thermal stability
(B) High fidelity
(C) High speed
(D) High thermal stability

90. Introduction of recombinant DNA into bacterial cell by using current is called
(A) Transformation
(B) Electroporation
(C) Transformation
(D) Tranduction

91. Nick translation is done by
(A) DNA polymerase I
(B) DNA polymerase II
(C) DNA ligase
(D) Kinase

92. Telomeric sequences are found in
(A) HAC (B) BAC
(C) YAC (D) PAC

93. Transposons are known for the ________
(A) Mobile genetic elements between homologous chromosomes
(B) Mobile genetic elements between non homologous chromosomes
(C) Non mobile genetic elements
(D) Recombinant genetic elements

94. ______ is a set of techniques used to adapt plants for specific needs or opportunities
(A) Plant Biotechnology
(B) Animal Biotechnology
(C) Nanobiotechnology
(D) Molecular Genetics

95. Plasmids replication depends on the specific factors and _______
(A) Autonomous
(B) Depends on the genomic DNA
(C) Needs an initiator
(D) Does not replicate

96. Ti plasmid is useful in bringing _______
(A) New genes into animal cells
(B) New genes into plant cells
(C) Tumor cells into plant cells
(D) Tumor cells into animal cells

97. ______ is a set of techniques used to adapt plants for specific needs or opportunities
(A) Plant Biotechnology
(B) Animal Biotechnology
(C) Nanobiotechnology
(D) Molecular Genetics

98. Plasmids replication depends on the specific factors and _______
(A) Autonomous
(B) Depends on the genomic DNA
(C) Needs an initiator
(D) Does not replicate

99. Ti plasmid is useful in bringing ______
(A) New genes into animal cells
(B) New genes into plant cells
(C) Tumor cells into plant cells
(D) Tumor cells into animal cells

100. Ti plasmid transfer works with______
(A) Monocots only
(B) Dicots only
(C) All plants
(D) Both monocots and dicots

101. Ti plasmid naturally occurs in ______
(A) Agrobacterium tumefaciens
(B) Corynebacterium species
(C) Staphylococcus species
(D) Vibrio chloreae

102. Plasmids are naturally occurring as ______
(A) Linear single stranded DNA
(B) Linear single stranded RNA
(C) Linear duplex DNA
(D) Circular duplex DNA

103. The plasmids associated with resistance to antibiotics is represented as ______
(A) R (B) Hfr
(C) D (D) F

104. Plasmids with ability to coexist in the same cell with other plasmids are called ______
(A) Major (B) Classical
(C) Incompatibility (D) Constitutive

105. Small pieces of DNA that can insert themselves into chromosomes at numerous locations and mutations are called ______
(A) Carcinogens
(B) Wild type genes
(C) Transposons
(D) Frame shift genes

106. The plasmid resist to ampicillin is ______
(A) pBR327 (B) pBR322
(C) pUC (D) pCAMBIA

107. The plasmid derived from *E. coli* is ______
(A) pBR327 (B) pBR322
(C) plasmid R1 (D) pCAMBIA

108. pBR322 vector was produced by ______
(A) Bolivar (B) Rodriguez
(C) Both A and B (D) Landsteiner

109. Antisense RNA molecules have a sequence ______ to normal RNA transcripts
(A) Complementary
(B) Non complementary
(C) Opposite
(D) Similar

110. Gene silencing is generally termed as ______ of genes
(A) Switching off (B) Switching on
(C) Absence (D) Presence

111. Production of ethylene is inhibited by antisense gene______
(A) Glyphoshate
(B) ACC synthase
(C) ACC synthatase
(D) Lyase

112. ______ is responsible for fruit ripening
(A) Glyphoshate
(B) ACC synthase
(C) ACC synthatase
(D) Polygalacturonidase

113. Which one of the following technique is used to produce Flavr Savr tomato?
(A) rDNA technology
(B) Trangenesis
(C) Antisense RNA technology
(D) rRNA technology

114. Preservation of germplasm in the frozen state is termed as ______
(A) Cryoprectectant
(B) Cryopreservation
(C) Preservation
(D) Storage

115. What is a cloning vector?

(A) An agent used to transfer DNA from an in vitro solution into a living cell

(B) The sticky end of a DNA fragment

(C) The laboratory apparatus used to clone genes

(D) A DNA probe used to locate a particular gene in the genome

116. What are the essential characteristics of a cloning vector?

(A) Bacterial cells cannot survive without it

(B) Bacterial cells replicate it

(C) Bacterial cells take it up

(D) Both b and c

117. Which bacteria are known as natural genetic engineer of plants?

(A) *Agrobacterium tumefaciens*

(B) *E. coli*

(C) *Bacillus* sp

(D) *Streptomyces* sp

118. Who coined the term plasmid?

(A) Herbert Boyer

(B) Lederberg

(C) Stanley

(D) Benthem

119. A transgenic organism has _______

(A) Genes that other organisms do not

(B) Genes from different species

(C) A dominant phenotype

(D) Had its genome sequenced

120. Chromosome tips, or ____, function as a cellular clock that ticks down as pieces are lost from the very ends

(A) Chromatids (B) Centromeres

(C) Centrioles (D) Telomeres

121. What is the meaning of "+" or "-" in the name of the cloning vector pBluescript SK (+/-)?

(A) Describes whether sense or antisense transcripts will be produced from the T7 promoter

(B) Describes the orientation of the multiple cloning site

(C) Means two different quality control methods applied to the product

(D) Describes the orientation of the f1 phage ori

122. What is a chimeraplast?

(A) An RNA/DNA hybrid oligonucleotide

(B) A fusion product of protoplasts from different species

(C) A transformed plastid carrying a transgene

(D) A product of 1st strand cDNA synthesis

123. What is essential for T-DNA transfer?

(A) Inverted repeat border sequences

(B) Palindromic border sequences

(C) Direct repeat border sequences

(D) An AtuI restriction site at the border sequence

124. What are the precursors for the synthesis of nopalines?

(A) Phosphoenolpyruvate + amino acid

(B) Pyruvate + amino acid

(C) Alpha-Ketoglutarate + amino acid

(D) Oxalate + amino acid

125. Manufacturing recombinant DNA molecules involves cutting a gene from its normal location, inserting it into a circular piece of DNA from a bacterial cell, and then transferring the circle of DNA to cells of another species. Which of the below describe the circular piece of DNA from a bacterial cell?

(A) Restriction enzyme

(B) Plasmid

(C) Bacteriophage

(D) Vector

126. Which gene transfer technique involves a tiny needle which is used to inject DNA into a cell lacking that DNA sequence?

(A) Electroporation
(B) Liposome transfer
(C) Microinjection
(D) Particle bombardment

127. The process of ____ involves the introduction of a gene into a cell where it exchanges places with its counterpart in the host cell

(A) Transgenic technology
(B) Gene targeting
(C) Knockout technology
(D) Recombinant DNA technology

128. Which gene transfer technique involves the use of a fatty bubble to carry a gene into a somatic cell?

(A) Electroporation
(B) Liposome transfer
(C) Microinjection
(D) Particle bombardment

129. _______ is to introduce a new trait to the plant which does not occur naturally in the species

(A) Genetically modified crops
(B) GMCs
(C) Biotech crops
(D) GMOs

130. Which one of the following process is used for the fusion of a normal protoplast with an enucleated protoplast?

(A) Cybridization
(B) rDNA technology
(C) Transgenesis
(D) Hybridization

131. Cytoplasmic hybrids refer to________

(A) Cytoplast (B) Chloroplast
(C) Tonoplast (D) Protoplast

132. The source of cellulolysin is _______

(A) *Trichoderma viride*
(B) *Irpex lactus*
(C) *Aspergillus* species
(D) *Rhizopus* species

133. The source of Helicase is _______

(A) *Trichoderma viride*
(B) *Irpex lactus*
(C) *Helix pomatia*
(D) *Rhizopus* species

134. The source of Zymolyase is _______

(A) *Trichoderma viride*
(B) *Arthrobacter luteus*
(C) *Helix pomatia*
(D) *Rhizopus species*

135. FDA stands for _______

(A) Fluorescein diacetate
(B) Federal diacetate
(C) Fluorescein diacetate assay
(D) Federal Development Assay

136. The transfer of desirable genes from one plant species to another is called as ______ technology

(A) rDNA
(B) Transgenic
(C) Plant genetic transformation
(D) Plant transformation

137. What is the length of Ti plasmid?

(A) 20kb (B) 200kb
(C) 2kb (D) 2000kb

138. Which one of the following technique is mostly confined to protoplasts that can be regenerated to viable plants?

(A) Electroporation
(B) Biolistic method
(C) Microinjection
(D) Silicon carbide fibres

139. Opines are _______

(A) Aminoacids (B) Lipids
(C) Proteins (D) Nucleic acids

140. T-DNA borders are a set of _____ kb sequences present on either side of TDNA

(A) 25 (B) 23
(C) 24 (D) 22

141. Which of the following region codes for proteins involved in the uptake and metabolism of opines in Ti plasmids?

(A) Promoter
(B) Vir region
(C) Marker region
(D) Opine catabolism region

142. Which one of the following plants have been developed for expressing antigens derived from animal viruses?

(A) Tomato (B) Tobaco
(C) Potato (D) Corn

143. The first clinical trials in humans, using a plant derived vaccine were conducted in _____

(A) 1987 (B) 1997
(C) 1977 (D) 1967

144. The first field tests were of what genetically altered organism?

(A) bt corn
(B) Vaccinia virus containing a gene from the rabies virus
(C) Strawberry seedlings sprayed with "ice-minus" bacteria
(D) The "flavrsavr" tomato

145. In a ___________ protocol, bacteria with engineered abilities to detoxify pollutants are intentionally released in an area

(A) Microcosm establishment
(B) Bioremediation
(C) Mibridization
(D) Rhizosecretion

146. Which of the following dies from Ti plasmid infection?

(A) Grains (B) Banana
(C) Palm (D) Rice

147. Which of the agricultural challenges below cannot be solved with transgenic techniques?

(A) Public preference for organic vegetables
(B) Crops are killed by a virus
(C) Public concern about safety of synthetic pesticides
(D) Crops are damaged by frost

148. There is a restriction endonuclease called EcoRI. What does "co" part in it stand for?

(A) Coli (B) Colon
(C) Coelom (D) Coenzyme

149. The most common substrate used in distilleries for the production of ethanol is

(A) Molasses (B) Corn meal
(C) Soya meal (D) Ground gram

150. Which of the following is not tool of genetic engineering?

(A) Vectors (B) Enzymes
(C) Foreign DNA (D) GMO

151. In recombinant DNA technology a plasmid vector is cleaved by

(A) Modified DNA ligase
(B) A heated alkaline solution
(C) The same enzyme that cleave the donor DNA
(D) The different enzyme other than that cleave the donor DNA

152. The most common plasmid vector used in genetic engineering is

(A) PBR 328 (B) PBR 322
(C) PBR 325 (D) PBR 330

153. 'Nif gene' for nitrogen fixation is cereal crops like wheat, jowar etc. is introduced by cloning

(A) *Rhizobium meliloti*
(B) *Bacillus thuringiensis*
(C) *Rhizopus*
(D) *Rhizophora*

154. EcoR1 is an
(A) Ligase
(B) Polymerase
(C) Restriction enzyme
(D) Gyrase

155. The first transgenic crop was
(A) Pea (B) Tobacco
(C) Flax (D) Cotton

156. Hirudin is obtained from the transgenic plant
(A) *Brassica napus*
(B) *Hibiscus rosasinesis*
(C) *Raphanus sativus*
(D) *Vinca rosea*

157. Bt Cotton is
(A) Cloned plant
(B) Transgenic plant
(C) Hybrid plant
(D) Mutated plant

158. Dolly sheep was genetically similar to
(A) The mother from which nucleated fertilized egg was taken
(B) The mother from which nuclear DNA of udder cell was taken
(C) The surrogate mother
(D) Both surrogate mother and nuclear donor mother

159. Which organism was used as bioweapon derived from______
(A) Clostridium
(B) *Yerstsinia pestis*
(C) *Fusarium* species
(D) Green algae

160. A set standards used to regulate own or community activity in relation to biological world is
(A) Biopotency (B) Biopiracy
(C) Biowar (D) Bioethics

161. Biopiracy means
(A) Use of biopatents
(B) Thefts of plants and animals
(C) Stealing of bioresources
(D) Exploitation of bioresources without authentic permission

162. The first hormone artificially produced by bacteria is
(A) Insulin (B) Thyroxin
(C) Adrenalin (D) Testosterone

163. In which crop, the first transgenic was developed?
(A) Tomato (B) Soybean
(C) Tobacco (D) Cotton

ANSWER SHEET

1. **C**	2. **B**	3. **B**	4. **D**	5. **C**	6. **D**	7. **C**	8. **A**	9. **B**	10. **A**
11. **D**	12. **B**	13. **A**	14. **C**	15. **D**	16. **B**	17. **C**	18. **A**	19. **B**	20. **A**
21. **C**	22. **A**	23. **C**	24. **D**	25. **D**	26. **C**	27. **C**	28. **D**	29. **A**	30. **A**
31. **C**	32. **D**	33. **D**	34. **C**	35. **C**	36. **A**	37. **C**	38. **A**	39. **A**	40. **C**
41. **D**	42. **A**	43. **A**	44. **A**	45. **B**	46. **B**	47. **C**	48. **B**	49. **B**	50. **D**
51. **C**	52. **A**	53. **D**	54. **B**	55. **D**	56. **D**	57. **B**	58. **A**	59. **D**	60. **C**
61. **C**	62. **C**	63. **C**	64. **A**	65. **A**	66. **D**	67. **B**	68. **B**	69. **B**	70. **D**
71. **B**	72. **A**	73. **C**	74. **A**	75. **C**	76. **C**	77. **D**	78. **B**	79. **B**	80. **D**

81. **C**	82. **C**	83. **B**	84. **C**	85. **D**	86. **B**	87. **D**	88. **D**	89. **D**	90. **B**
91. **A**	92. **C**	93. **B**	94. **A**	95. **A**	96. **C**	97. **A**	98. **A**	99. **B**	100. **B**
101. **A**	102. **C**	103. **A**	104. **C**	105. **C**	106. **D**	107. **C**	108. **C**	109. **A**	110. **A**
111. **B**	112. **D**	113. **C**	114. **A**	115. **A**	116. **A**	117. **A**	118. **B**	119. **B**	120. **D**
121. **D**	122. **A**	123. **B**	124. **C**	125. **B**	126. **D**	127. **A**	128. **D**	129. **A**	130. **D**
131. **C**	132. **B**	133. **A**	134. **C**	135. **D**	136. **A**	137. **B**	138. **B**	139. **C**	140. **D**
141. **D**	142. **A**	143. **B**	144. **C**	145. **B**	146. **D**	147. **A**	148. **A**	149. **A**	150. **D**
151. **C**	152. **B**	153. **A**	154. **C**	155. **B**	156. **A**	157. **B**	158. **B**	159. **C**	160. **D**
161. **D**	162. **A**	163. **C**							

Chapter 4

GENOMICS

1. The bread wheat, *Triticum aestivum*, that is commonly used all over the world, is
   (A) Hexaploid (B) Tetraploid
   (C) Triploid (D) Diploid
2. The difference between a simple tandem array and a compound tandem array is
   (A) The frequency of different sequences
   (B) The direction the sequences face (left or right)
   (C) The number of different sequences involved
   (D) The number of repetitions of sequences
3. In the creation of a microarray (biochip), fragments of DNA are placed on a
   (A) Microscope slide
   (B) Circuit board
   (C) Silicone chip
   (D) Membrane
4. Proteomics is the study of
   (A) All proteins in an organism
   (B) Gene-transcribing proteins
   (C) Structural proteins of a cell
   (D) Regulating proteins in humans
5. Increased genetic diversity following extended time in a tissue culture is a problem called
   (A) Gene alteration
   (B) Temporal modification
   (C) Somaclonal variation
   (D) Culture shock
6. A mass of dividing, undifferentiated cells in a tissue culture is called
   (A) A shield (B) A callus
   (C) An embryoid (D) An aggregate
7. Approximately ________ % of the world's production of fats and oils comes from plants
   (A) 15 (B) 25
   (C) 45 (D) 75
8. Electroporation is a technique used with
   (A) Calluses (B) Pollen
   (C) Protoplasts (D) Organs
9. To produce plants that are homozygous for all traits, the best choice is
   (A) Cell suspension culture
   (B) Callus culture
   (C) Anther / pollen culture
   (D) Plant organ culture
10. Chloroplast DNA is
    (A) Separate from nuclear DNA
    (B) Coded by the nucleus
    (C) Paternally inherited
    (D) A subset of nuclear DNA
11. Cell suspension cultures require
    (A) Organogenesis
    (B) Electroporation
    (C) Differentiation
    (D) Disaggregation

12. The function of "high-copy-number DNA" is
(A) Encoding most genes
(B) Encoding rRNA
(C) Encoding a single gene
(D) Unknown

13. Genetic engineering has modified _______ for the creation of mono-unsaturated fatty acids
(A) PCR denaturase
(B) QTL polymerase
(C) LCP lipase
(D)ACP desaturase

14. Genetic engineering has succeeded in modifying canola oil to replace
(A) Yogurt (B) Coconut oil
(C) Palm oil (D) Cocoa butter

15. Most crops grown in the United States yield less than _______________ % oftheir genetic potential
(A) 10 (B) 20
(C) 50 (D) 60

16. If you want to use a plant tissue culture as a chemical factory for vitamins, choose
(A) Suspension cultures
(B) Callus cultures
(C) Organ cultures
(D) Anther pollen cultures

17. The function of "low-copy-number DNA" is
(A) Encoding most genes
(B) Encoding rRNA
(C) Encoding a single gene
(D) Encoding tRNA

18. The fastest way to a ripe tomato using tissue culture is by
(A) Anther / pollen culture
(B) Plant organ culture
(C) Protoplast culture
(D) Callus culture

19. Characteristic feature of the chloroplast genome is the presence of
(A) Two simple tandem arrays
(B) Two repeat / repeat interspersions
(C) Two identical inverted repeats
(D) Two compound tandem arrays

20. Tools to detect polymorphism in plants are _______________ maps
(A) RFLP and QTL
(B) AFLP and PCR
(C) PCR and QTL
(D) RFLP and AFLP

21. A plant useful in the study of plant genomes because of its small genome size is
(A) *Avena sativa*
(B) *Petunia porodii*
(C) *Tulipa* spp
(D) *Arabidopsis thaliana*

22. Sequencing efforts are underway for ____________ as a model for a small monocot genome
(A) Rice (B) Corn
(C) Barley (D) Wheat

23. Most plant tissue cultures are initiated from
(A) Calluses (B) Explants
(C) Plantlets (D) Protoplasts

24. Pores in protoplasts may be opened to DNA by the application of
(A) Magnetism
(B) Light
(C) Enzymes
(D) Electricity

25. A large inversion of chloroplast DNA is found in the __________ family
(A) Castor bean
(B) Sunflower
(C) Oats
(D) Arabidopsis

26. The function of "medium -copy -number DNA" is

(A) Encoding most genes
(B) Encoding rRNA
(C) Encoding a single gene
(D) Encoding tRNA

27. The high-copy-number-DNA in plants

(A) Codes for rRNA
(B) Codes for proteins required for plant metabolism
(C) Are called transposable elements
(D) Have no known function

28. Which of the following are mismatched?

(A) Simple tandem array-repetitive sequence in the same orientation
(B) Compound tandem array-repetitive sequence in the same orientation
(C) Inverted repeats-repetitive sequence in the opposite orientation
(D) Single-copy gene-sequence is not repeated

29. The DNA in chloroplasts

(A) Is more complex than nuclear DNA
(B) Does not have prokaryotic features
(C) Is subject to modifications caused by transposable elements
(D) Is less diverse than the nuclear DNA in plants

30. The genome sequencing of rice is important because

(A) The rice genome is very unique and contains genes not found in other plants
(B) The rice genome is very large compared to the DNA of other grains and so more genes will be identified
(C) It can identify genes associated with disease resistance, growth capacity, etc
(D) It has a rapid life cycle and has many identifiable mutations

31. Unlike the somatic cells of animals, plant cells can be grown in tissue culture and regenerate new plants

(A) Because each cell contains the entire genome
(B) Because plant cells can dedifferentiate and revert back to seeds
(C) Because plant cells are able to express genes that weren't previously expressed
(D) Both A and B

32. Which of the following are mismatched?

(A) Cell suspension culture-somatic cell embryogenesis
(B) Protoplast isolation-hybrids
(C) Homozygous plants-anther/pollen culture
(D) Callus culture-differentiated tissues regenerate

33. Plant transformation is

(A) When a plant grown in culture generates increased genetic variation
(B) When plant cells in suspension cultures form individual embryos that can grow into plants
(C) The incorporation of foreign DNA into the plant genome
(D) When dedifferentiated callus cells develop into tissue that is different from the original source tissue

34. The collection of proteins that can be produced by a given species is:

(A) Considered that species' genetic complement
(B) Correlated with the size of the organism
(C) Called the proteome
(D) Called the transcriptome

35. VNTRs represnets-

(A) New terminal regions in DNA
(B) Functional genes in the DNA
(C) Split genes in the sample DNA
(D) Specific non-coding sequences with unique tandem repeats

36. DNA fingerprinting is based on
(A) Occurance of VNTR's
(B) Knowledge of human karyotype
(C) Cloned DNA
(D) Recombinant DNA

37. In forensic science which of the following is used?
(A) Bacterial cloning
(B) DNA foot printing
(C) DNA fingerprinting
(D) DNA cloning

38. Which of these would not be a valid reason that use of microarray technology to differentiate between closely related bacterial species and subspecies is important?
(A) Certain strains of bacteria are more pathogenic than other related strains
(B) Some strains of bacteria are more active in bioremediation than other related strains
(C) Infection by different strains of bacteria may require different therapeutic approaches
(D) In many cases, critical information about characteristics of a bacterium causing an infection needs to be immediately available

39. The two most common processes that lead to production of multiple functional proteins from the same DNA sequence are:
(A) RNA editing and alternative splicing
(B) Protein folding and posttranslational covalent modifications
(C) Alternative splicing and posttranslational covalent modifications
(D) Post-translational covalent modification and transcriptional regulation

40. Which of the following statements regarding the proteome is the most correct?
(A) A large proportion of the proteome is expressed by each cell of a species
(B) Levels of gene expression at the mRNA level generally correlate highly with levels of functional protein
(C) The proteins produced by a specific cell depend on cell type and environmental conditions
(D) All of the above

41. Sequencing of genomic DNA is included under
(A) Structural genomics
(B) Functional genomics
(C) Proteomics
(D) Transgenesis

42. Gene expression, regulation and phenotype production are studied in second phase of genome analysis called
(A) Structural genomics
(B) Functional genomics
(C) Proteomics
(D) Transmeiosis

43. A flowering plant lily have ______ more DNA than humans
(A) 10 times (B) 15 times
(C) 18 times (D) 13 times

44. Which of these concerns would apply to functional protein microarrays but not antibody microarrays?
(A) Antibodies recognize relatively short peptide sequences from a protein, and presence of the same peptide sequence in several proteins is possible
(B) Three dimensional structure of a protein is essential for correct function, and the process of microarray preparation may interfere with this structure
(C) Certain functions, such as structural support of a cell, may not be amenable to analysis on a microarray
(D) All of these apply to functional protein microarrays but not antibody microarrays

45. Which of the following pieces of information would not be required for a computer program to determine exon/intron structure of a gene?

(A) A table correlating codons to amino acids
(B) The nucleotide sequence of the mRNa
(C) Splice site signals for the species under study
(D) All of these pieces of information would be required

46. How many potential open reading frames are present in a DNA sequence?

(A) One (B) Three
(C) Six (D) More than six

47. You have cloned a new gene from your favorite species of experimental organism, which has had very few genes cloned so far. Which of the following techniques would you expect to provide information about the function of this gene?

(A) Comparison of regulatory sequences to other DNA sequences in this organism
(B) Comparison of regulatory sequences to DNA sequences from other organisms
(C) Comparison of the gene sequence to other DNA sequences in this organism
(D) Comparison of the gene sequence to DNA sequences from other organisms

48. Tools to detect polymorphism in plants are ______________ maps

(A) RFLP, AFLP (B) AFLP, PCR
(C) PCR, QTL (D) RFLP, QTL

49. A plant useful in the study of plant genomes because of its small genome size is _______

(A) *Avena sativa*
(B) *Petunia porodii*
(C) *Tulipa* sp
(D) *Arabidopsis thaliana*

50. Sequencing efforts are underway for ______________ as a model for a small monocot genome

(A) Rice (B) Corn
(C) Barley (D) Wheat

51. Genome is

(A) Genes on nuclear DNA
(B) Nuclear DNA + mitochondrial DNA
(C) Nuclear DNA + chloroplast DNA
(D) Nuclear DNA + Mitochondrial DNA + Chloroplast DNA

52. A technique of using very small metal particles coated with desired gene in the gene transfer is called

(A) Electroporation
(B) Microinjection
(C) Liposome
(D) Biolistics

53. The complete set of chromosomal and extrachromosomal genes of an organisms is called

(A) Genome (B) Gene pool
(C) Gene bank (D) Gene library

54. The study of all the proteins coded by the genome is called

(A) Proteome (B) Proteomics
(C) Genome (D) Protein formation

55. Genomics is the study of genomes. Genome refers to the

(A) DNA of an organism
(B) Total DNA and RNA of an organism
(C) Entire genes of an organism
(D) Total DNA, RNA and cDNA of an organism

56. The term genomics was coined by

(A) Thomas Cech
(B) T.H Morgan
(C) Thomas Roder
(D) Craig Venter

57. The first completed genome sequencing project is of

(A) *E. coli*
(B) *Haemophilus influezae*
(C) ØX174
(D) *Drosophila melanogaster*

58. International Human Genome project was initiated by

(A) National Institute of Health (NIH)
(B) Celera genomics
(C) US Department of Energy (DoE)
(D) NOH and US DoE

59. DNA sequencing followed by genome annotation are steps of

(A) Comparative genomics
(B) Structural genomics
(C) Functional genomics
(D) Transcriptomics

60. All are genome sequencing strategies except

(A) Edman degradation method
(B) Short gun library
(C) Whole genome short gun sequencing
(D) Directed gene sequencing

61. Small cDNA sequence that represents a unique segment of an

active gene is called

(A) SNPs (B) SnRNAs
(C) ESTs (D) Contigs

62. Which of the following is not a gene expression database?

(A) GenBank (B) Flyview
(C) SeedGenes (D) BodyMap

63. Variation between individuals due to single base changes is called as

(A) ESTs (B) Contigs
(C) SNPs (D) Transversion

64. Alignment method suitable for aligning closely related sequence is

(A) Multiple sequence alignment
(B) Pair wise alignment
(C) Global alignment
(D) Local alignment

65. The alignment procedure that tries to align the entire sequence is

(A) Multiple sequence alignment
(B) Pair wise alignment
(C) Global alignment
(D) Local alignment

66. The alignment method suitable for finding out conserved patterns in DNA or protein sequences is

(A) Multiple sequence alignment
(B) Pair wise alignment
(C) Global alignment
(D) Local alignment

67. The procedure of aligning many sequences simultaneously is called

(A) Multiple sequence alignment
(B) Pair wise alignment
(C) Global alignment
(D) Local alignment

68. _______ marker is a DNA sequence in the genome which can be located and identified

(A) Molecular (B) Genetic
(C) Biochemical (D) Physical

69. The alignment procedure that tries to align regions with high level of matches without considering the alignment of rest of the sequences is

(A) Multiple sequence alignment
(B) Pair wise alignment
(C) Global alignment
(D) Local alignment

70. All are sequence alignment tools except

(A) Rasmol (B) BLAST
(C) FASTA (D) Clustal W

71. Which of the following is a sequence alignment tool provided by NCBI
(A) Chime (B) BLAST
(C) FASTA (D) Clustal W

72. Which of the following is a multiple sequence alignment tool?
(A) Clustal W (B) Chime
(C) Dismol (D) PDB

73. What type of products is generated after RAPD?
(A) Orbitory (B) Arbitrary
(C) Auxillary (D) Orbitol

74. LINES stand for ______ Nuclear Elements
(A) Long Interspersed
(B) Large Interspersed
(C) Long Interpierced
(D) Large Interpierced

75. What are the two types of classifications under Long Terminal Repeats in retroposons?
(A) Ty1, Ty2 (B) Ty1, Ty4
(C) Ty1, Ty3 (D) Ty2, Ty3

76. How much percentage of human genome is composed of transposons?
(A) 50 (B) 40
(C) 30 (D) 20

77. What is class I type transposons?
(A) Transposons III
(B) Transposons I
(C) Transposons II
(D) Retroposons

78. RAPD is also known as ______
(A) Orbitory primer PCR
(B) Orbitol primer PCR
(C) Arbitrary primer PCR
(D) RFLP

79. RAPD is a molecular marker based on ______ amplification
(A) RFLP (B) PCR
(C) AFLP (D) SSR

80. In RAPD amplification product will be generated only between ______ primer annealing sites
(A) 3 (B) 2
(C) 4 (D) 5

81. Who postulated the existence of genetic elements which transpose from one location in the genome to another?
(A) Larkin
(B) Evans
(C) Barbara Mcclintock
(D) Bidney

82. Which one of the following serves as a marker for mapping a specified gene in RFLP?
(A) DNA (B) rRNA
(C) tRNA (D) mRNA

83. A variant in sequence that is present in at least 1% of a population is called a (an) ______
(A) Gene (B) Allele
(C) Autosome (D) Polymorphism

84. A gene may have many alleles, but each individual has only two alleles because ______
(A) Having more than two alleles is lethal
(B) Having more than two alleles unbalances the chromosomes
(C) A person has two parents, each will contribute one allele
(D) A backup set of alleles is necessary in case something

85. DNA finger printing was developed by
(A) Francis Crick (B) Khorana
(C) Alec Jeffrey (D) James Watson

86. The technique to distinguish the individuals based on their DNA print patterns is called

(A) DNA fingerprinting

(B) DNA profiling

(C) Molecular fingerprinting

(D) All of these

87. DNA finger printing relies on

(A) Difference in patterns of genes between individuals

(B) Difference in order of genes between individuals

(C) Difference in junk DNA patterns between individuals

(D) All of these

88. Minisatellites are

(A) 10-40 bp sized short sequences within the genes

(B) Short coding repetitive regions on the eukaryotic genome

(C) Short Non-coding repetitive sequences present throughout the chromosome

(D) Are regions of chromosomes after secondary constriction

89. DNA profiling technique to demonstrate the similarity between different animal species with reference to some specific protein coding DNA sequences is called

(A) Zoo blot

(B) Phylogenetic blot

(C) Animal profiling

(D) Animal blot

90. The DNA fingerprint pattern of a child is

(A) Exactly similar to that of both of the parents

(B) 100% similar to the father's DNA print

(C) 100% similar to the mother's DNA print

(D) 50% bands similar to father and rest similar to

91. Specific biomolecules which show easily detectable differences among different strains of a species or among different species is termed as

(A) DNA fingerprinting

(B) Molecular markers

(C) Molecular scissors

(D) RFLP

92. The variation in the restriction DNA fragment lengths between individuals of a species is called

(A) Restriction Fragment Length Polymorphism (RFLP)

(B) Random amplified Polymorphic

(C) Amplified Fragment Length Polymorphism (AFLP)

(D) Simple Sequence repeats (SSR)

93. RFLP involves

(A) Used to identify a specific protein

(B) Used to identify a specific DNA

(C) Used to identify a specific RNA

(D) Used to identify both DNA and RNA RFLP

94. The set of DNAs generated by using random primers in a PCR reaction is called

(A) RAPD

(B) RFLP

(C) AFLP

(D) *In situ* hybridization

95. All the statements are true regarding RFLP and RAPD except

(A) RAPD is a quick method compared to RFLP

(B) RFLP is more reliable than RAPD

(C) Species specific primers are required for RAPD

(D) Radioactive probes are not required in RAPD

96. The variation in number of tandem repeats between two or more individuals is called

(A) Variable number of tandem repeats (VNTRs)

(B) Restriction Fragment Length Polymorphism (RFLP)

(C) Simple sequence repeats (SSRs)

(D) Amplified Fragment Length Polymorphism (AFLP)

ANSWER SHEET

1. **A**	2. **C**	3. **A**	4. **A**	5. **C**	6. **B**	7. **D**	8. **C**	9. **C**	10. **A**
11. **D**	12. **D**	13. **D**	14. **D**	15. **C**	16. **A**	17. **A**	18. **B**	19. **C**	20. **D**
21. **D**	22. **A**	23. **B**	24. **D**	25. **B**	26. **B**	27. **D**	28. **B**	29. **D**	30. **C**
31. **C**	32. **C**	33. **C**	34. **C**	35. **D**	36. **A**	37. **C**	38. **D**	39. **C**	40. **C**
41. **A**	42. **B**	43. **B**	44. **C**	45. **B**	46. **C**	47. **D**	48. **A**	49. **D**	50. **A**
51. **D**	52. **D**	53. **A**	54. **B**	55. **C**	56. **C**	57. **C**	58. **D**	59. **B**	60. **A**
61. **C**	62. **A**	63. **C**	64. **C**	65. **C**	66. **D**	67. **A**	68. **A**	69. **D**	70. **A**
71. **B**	72. **A**	73. **B**	74. **A**	75. **C**	76. **A**	77. **D**	78. **D**	79. **B**	80. **B**
81. **C**	82. **A**	83. **D**	84. **C**	85. **C**	86. **D**	87. **C**	88. **C**	89. **A**	90. **D**
91. **B**	92. **A**	93. **C**	94. **A**	95. **C**	96. **A**				

Chapter 5

PLANT TISSUE CULTURE

1. Hormone pair required for a callus to differentiate are
   (A) Auxin and ABA
   (B) Auxin and Cytokinin
   (C) GA and Cytokinin
   (D) GA and Ethylene

2. The term mineral elements was proposed by
   (A) Calvin (B) Stephan Hales
   (C) Godlewski (D) Arnon and Stout

3. Which TC banana cultivar commercially available in the world
   (A) Pome group –Hill banana (AAB)
   (B) Dessert group –Poovan (AAB)
   (C) Plantain group-Nendran(AAB)
   (D) Cavendish group- Grand Naine(AAA)

4. The term somaclonal variation was coined by
   (A) Darwin (B) Mendel
   (C) Morgan
   (D) Larkin and Scowcroft

5. Which of the following plant cell will show totipotency
   (A) Sieve tube (B) Pith cell
   (C) Cork Cell (D) Meristem

6. Optimum temperature for enzyme activity in protoplast isolation
   (A) 10-20 (B) 90 -120
   (C) 40-50 (D) 0-4

7. Commonly using antioxidant in plant tissue culture
   (A) Acidic acid (B) Citric acid
   (C) Ferric acid (D) HCL

8. Cassava mosaic virus is
   (A) ssRNA (B) ssDNA
   (C) dsRNA (D) dsDNA

9. The first anther culture in Datura was carried out by
   (A) Beadle and Tatum
   (B) Swaminathan
   (C) Guha and Maheswarid
   (D) Linsmaier and Skoog

10. The variations observed in the regenerated plants from gametic cells
    (A) Clonal variations
    (B) Gamotoclonal variations
    (C) somaclonal variation
    (D) variations

11. The average pH of the double distilled water is
    (A) 4 (B) 8.5
    (C) 0.0 (D) 7

12. Hand book on plant tissue culture was written by
    (A) Gautheret (B) Kanta
    (C) Skoog & Miller (D) Morel

13. The first attempt in developing haploid Datura plants using pollen grains was proposed by

(A) Maheswari & Scowcroft

(B) Guha & Maheswari

(C) Murry & Skoog

(D) Murashige

14. How much grams of agar per litre of medium is satisfactory?

(A) 6 to 8 gms (B) 4 to 6 gms

(C) 0.8 to 1.0 gms (D) 1.0 to 1.2 gms

15. Isopentenyl adenine is an example for

(A) Natural cytokinin

(B) Artificial cytokinin

(C) Auxin

(D) Gibberellin

16. In a tissue culture technique when a plant is grown in culture medium certain macro & microelement are provided as nutrient. One of these element is

(A) Leusine (B) Alanine

(C) Glycine (D) Aspartic acid

17. Sucrose is generally used of the concentration of

(A) 3 to 6% (B) 2 to 5%

(C) 1 to 4% (D) 3 to 8%

18. Dry sterilization is carried out for

(A) 1 hr at 120-140°C

(B) 2 hrs at 160- 180°C

(C) 3 hrs at 160- 180°C

(D) 4 hrs 120- 140°C

19. Among the given surface sterilizing agents, which one gives good results?

(A) Calcium/sodium hypochloride

(B) Cetrimide

(C) Silver nitrate

(D) Bromine water

20. The plant culture taken for development of embryo from somatic tissues for the first time was by using

(A) *Datura carota*

(B) *Beta vulgaris*

(C) *Abelmoschus*

(D) *Anacardium occidendale*

21. The culture of whole plants and organ has been termed as

(A) Organized culture

(B) Non-organized culture

(C) Cell culture

(D) Protoplast culture

22. Cartharanthus roseus provided a rich

(A) Terpenoids indole alkaloids

(B) Pyrethrins

(C) Rotenoids

(D) Quinine

23. A culture is one in which cells are retained and inflow of fresh medium is balanced by outflow of corresponding volumes of spent medium only.

(A) Closed continuous

(B) Open continuous

(C) Batch

(D) Semi continuous

24. The secondary product 'Shikonin' is obtained from

(A) *Catharanthus*

(B) *Morinda*

(C) *Lithospermum erythrorhizon*

(D) *Papaver somaniferum*

25. How many pathways has been identified in the in vitro androgenesis?

(A) 1 (B) 2

(C) 3 (D) 4

26. The chromosome number of Hordeum vulgure is

(A) 2n = 7 (B) 2n = 14

(C) 2n = 21 (D) 2n = 28

27. The source pectolyase enzyme is

(A) *Aspergillus japonica*
(B) *Arthrobactor luteus*
(C) *Trichoderma viride*
(D) *Irpex lacteus*

28. The first somatic hybrid was produced by

(A) Skoog & coworkers
(B) Gamborg
(C) Coolson and coworkers
(D) Cocking workers

29. The variations arising in cell cultures of gametic origin like, in pollen and microspores cultures is known as

(A) Somaclonal (B) Gametoclonal
(C) Gynoclonal (D) Protoclonal

30. Protoplast was discovered by

(A) Mendel (B) Purkinje
(C) Darwin (D) Morgan

31. Multiplication of genetically identical copies of a cultivar by asexual reproduction is called________

(A) Clonal propagation
(B) Apomixis
(C) Vegetative propagation
(D) Micropropagation

32. The chemical which is used for surface sterilization of plants

(A) Silver Nitrate
(B) Sodium chloride
(C) Mercuric chloride
(D) Absolute alcohol

33. Callus is a

(A) Unorganized mass of cells
(B) Organized mass of cells
(C) Fully grown cells
(D) Differentiated cell

34. The hormone responsible for the callus induction

(A) Indole Acetic Acid
(B) Indole Butyric Acid
(C) 2,4- Dichloro phenoxy acetic acid
(D) Zeatin

35. The chemical which is used for diploidization is

(A) Colchicine
(B) Ethidium bromide
(C) Mercuric chloride
(D) Sodium hypochlorite

36. The ovary culture is most successful for the plants having

(A) Cytoplasmic male sterility (CMS)
(B) Self incompatibility
(C) Immature embryo
(D) Seed dormancy

37. The purpose of mature embryo culture is

(A) Embryo rescue
(B) Breaking seed dormancy
(C) Wide hybridization
(D) Avoid pre and post fertilization barriers

38. Who developed the technique micropropagation

(A) Morel & Martin
(B) White
(C) Skoog
(D) Power

39. For the complete eradication of Tobacco Mosaic Virus, the shoot tip culture is exposed to the temperature of

(A) 25-30 (B) 30-35
(C) 35-40 (D) 40-45

40. Which enzyme is not used in Protoplast isolation

(A) Cellulase
(B) Hemicellulase
(C) Pectinase
(D) Peroxidase

41. Which chemical is used for the production of Artificial or synthetic seed

(A) Sodium sulphate
(B) Sodium alginate
(C) Sodium molybdate
(D) Potassium chloride

42. Multiplication of plants through in vitro technique for commercial application is called----

(a) Micropropagation
(B) Macro production
(c) In vitro culture
(D) Tissue culture

43. Virus free plants can be obtained through

(a) Anther culture
(B) Meristem tip culture
(c) Callus culture
(D) Root tip culture

44. The first somatic hybrid was developed by

(a) Cocking (B) Power
(C) Carlson (D) Melcher

45. Which of the following statement is correct?

(a) Somaclonal variation are the caused by mutagens
(b) Somaclonal variation are the caused by gamma rays
(c) Somaclonal variation is produce during tissue culture
(d) Somaclonal variation induced during sexual embryogeny

46. Commonly using antioxidant in plant tissue culture

(A) Acidic acid
(B) Citric acid
(C) Ferric acid
(D) HCL

47. NCS-TCP strands for

(a) National Certification System for Tissue culture Plants
(b) National Certification System for Tree Plants
(c) National Central System for Tissue culture Plants
(d) National Certification System for Transgenic Plants

48. In cryopreservation, cells are stored in --------------

(A) Nitrogen (B) Helium
(C) Liquid nitrogen (D) Hydrogen

49. Optimum temperature for enzyme activity in protoplast isolation

(A) 10-20°C (B) 90 -120°C
(C) 40-50°C (D) 0-40°C

50. Virus free plants in Dahlia were first recovered by............

(A) Morel and Martin
(B) Morel and Morgan
(C) Murashiage & Skoog
(D) Martin & Skoog

51. Optimum pH of plant tissue culture MS medium............................

(A) 8.8 ± 0.2 (B) 6.8 ± 0.2
(C) 4.8 ± 0.2 (D) 5.8 ± 0.2

52. The commonly used carbon source in culture medium

(A) Sorbitol (B) Agarose
(C) Sucrose (D) Phytagar

53. Encapsulation of somatic embryos is called as

(A) Synthetic seed (B) Cybrids
(C) Vybrids (D) Hybrids

54. Variations observed during tissue culture are known as

(A) Clonal variations
(B) Somatic variations
(C) Somaclonal variations
(D) Tissue culture variations

55. Optimum temperature of plant growth room ____________

(A) (24+-1°C) (B) (4+-1°C)
(C) (60+-1°C) (D) (100+-1°C)

56. Example of commonly used gelling agent in plant tissue culture __________

(A) Sucrose (B) Agar Agar
(C) maltose (D) starch

57. Size of explants of meristem tip culture ____________

(A) 0.1 mm to 0.3 mm
(B) 2-3 m
(C) 1-4 cm
(D) 1000 cm

58. Chemical used for surface sterilization agent in Plant tissue culture _______

(A) 70% Ethanol (B) Sucrose
(C) CaCl2 (D) NaOH

59. __________ is important constitute of cell wall

(A) Na (B) P
(C) Ca (D) Fe

60. Enzyme used for protoplast isolation _____

(A) Restriction enzyme
(B) Nitrogenous enzymes
(C) Ligase
(D) Cellulase

61. Examples of wetting agent used in tissue culture ___________

(A) Ethanol (B) Tween 20
(C) NaoH (D) CaCl2

62. Commonly using carbon source in plant tissue culture

(A) Starch (B) Sucrose
(C) Mannitol (D) Fruit juice

63. Virus free plants can be produced by _____

(A) Callus culture
(B) Shoot tip culture
(C) Meristem tip culture
(D) Nodal culture

64. Ethylene is a ___________

(A) Growth promoter
(B) growth inhibitor
(C) weedicide
(D) enzyme

65. Nitrogen deficiency causes at _______

(A) Young leaf (B) New leaf
(C) Older leaf (D) Flower

66. Which medium using for tree micropropagation ____

(A) MS (B) WPM
(C) White (D) B5

67. Higher concentration of 2-4-D act as_____

(A) Growth hormone
(B) Herbicide
(C) Growth promoter
(D) Growth inhibitor

68. _______is the coding region of DNA

(A) Exon (B) Intron
(C) Promotor (D) Terminator

69. Optimum pH of plant tissue culture medium is _______

(A) 7 (B) 5.8
(C) 4 (D) 9

70. Organell DNA found in ________ system

(A) Virus (B) Animal and Plant
(C) Bacterium (D) Bacteriophage

71. Rooting powder contain ________

(A) BAP (B) GA
(C) ABA (D) IBA

72. ---------------------------- for the first time employed commercial enzyme preparation for isolation of protoplast

(A) Tulecke *et al* (B) Cocking *et al*
(C) Takabe *et al* (D) Skoog *et al*

73. Protoplast cultures are cultured at temperature ranging between----------------

(A) 20-25°C (B) 20-28°C
(C) 20-30°C (D) 18-28°C

74. ------------------- enzyme has both cellulolytic & pecteolytic activities

(A) Macerase (B) Diselase
(C) Helicase (D) pectinase

75. ---------------- pH of the enzyme used for protoplast isolation should ranged between

(A) 4.7 to 6.0 (B) 4.8 to 6.5
(C) 4.5 to 6.0 (D) 4.7 to 6.5

76. ----------------- proposed the production of synthetic seeds

(A) Zimmerman (B) Larkin
(C) Kohn (D) Kitto & Janick

77. ----------------- medium is developed for culture of soybean cell suspensions

(A) MS (B) LS
(C) B5 (D) N & N

78. ------------------ helps to increase the number of somatic embryos

(A) Alanine (B) Asparagine
(C) Cysteine (D) Serine

79. In general, callus cultures are subcultured after which of the following periods---------

(A) 4-6 days (B) 4-6 weeks
(C) 8-10 weeks (D) 2-3 months

80. In freeze preservation cell and tissues are preserved at ---------- in -----------

(A) -196°C Liquid nitrogen
(B) -193°C Liquid nitrogen
(C) -190°C Liquid carbon
(D) -190°C Liquid oxygen

81. ---------- is the most commonly used auxin for cell cultures

(A) 2,4-D (B) BAA
(C) PEG (D) Kinetin

82. ------------- is generally used as the carbon source in tissue culture media

(A) Sucrose (B) Maltose
(C) Fructose (D) Lactose

83. Ribavirin and Cyclohexinide act as --------------------- agents

(A) Antiviral (B) Antifungal
(C) Antimicrobial (D) Antibacterial

84. A localized group of actively dividing cells is known as--------------------

(A) Meristem (B) Stem
(C) Shoot (D) Root

85. A Somaclonal variety of B. juncea is -------------- which has been released for commercial cultivation as ' Pusa Jai Kisan'

(A) Varuna (B) Rober
(C) Ono (D) Pinder

86. Both Larkin and Scowcroft belong to which country---------------------

(A) America (B) Britain
(C) Australia (D) Germany

87. ____________ is the process of component cells of callus have the ability to form a whole plant

(A) Totipotency
(B) Redifferentiation
(C) Dedifferentiation
(D) Xylogenesis

88. The first report of interspecific hybridization through protoplast fusion on two species of Nicotiana was given by ______

(A) Carlson (B) Paul Berg
(C) Temin (D) Power

89. Shoot tip is a condensed shoot containing _________and leaf primordial with lateral bud

(A) Apical meristem
(B) Stem
(C) Young leaves
(D) Tendrils

90. A high endogenous ________level in shoot apices may inhibit the virus multiplication in meristem culture

(A) GA3 (B) Cytokinin
(C) Ethylene (D) Auxin

91. ________ is a Aromatic amino acid

(A) Alanine (B) Phenylalanine
(C) Valine (D) Leucine

92. The chemical used for protoplast fusion

(A) Polyethylene glycol
(B) Calcium hydroxide
(C) Dimethyl sulfoxide
(D) Potassium iodide

93. The peak accumulation of secondary metabolites occurs during

(A) Lag phase (B) Exponential pase
(C) Log Phase (D) Stationary phase

94. The White's medium was used for .... culture

(A) SP algae (B) Bacteria
(C) Fungi (D) Plant tissues

95. Commonly used Plant tissue culture media

(A) White's media (B) MS media
(C) B5 vitamins (D) knob's media

96. Among following growth harmones, which one is under cytokinin?

(A) IAA (B) BAP
(C) IBA (D) NAA

97. Evan's blue staining stains

(A) Dead cell (B) Live cell
(C) Viable protoplast
(D) Dividing cell

98. The optimal doubling time of plant cell is

(A) 1-2 hrs (B) 1-2 days
(C) 2-6 hrs (D) 2-6 days

99. Genomic DNA replication occur in

(A) G1 phase (B) M phase
(C) G2 phase (D) S phase

100. The ability of the component cells of callus to form a whole plant is known as

(A) Differentiation
(B) Redifferentiation
(C) Dedifferentiation
(D) Undifferentiation

101. Irradiation was measured by

(A) Joules (B) Lux
(C) Watt (D) Gray

102. The development of Haploids in anther culture was first reported by the Indian authors

(A) Gnanam and Kuzhanthaivelu
(B) Kumar et al
(C) Bhojwani and Razdan
(D) Guha and Maheswari

103. The best method of sterilizing hormones and vitamin solution is

(A) Autoclaving
(B) Dry air sterilization
(C) UV sterilization
(D) Filter sterilization

104. The main function of hormone Gibberlin is

(A) Root formation
(B) Shoot formation
(C) Ripening
(D) Internode elongation

105. The concentration of Potassium Nitrate in MS medium is

(A) 1650 mg/l
(B) 1900 mg/l
(C) 370 mg/l
(D) 440 mg/l

106. The citronella variety CIMAP/Bio13 was developed by which method

(A) Somaclonal variation
(B) Embryo rescue
(C) Protoplast fusion
(D) Callus culture

107. Father of micro- propagation

(A) E.A. Ball

(B) Toshio Murasighe and Folke Skoog

(C) Morel and Martin (1952)

(D) Gottlieb Haberlandt (1898)

108. Who did the work on virus free plants can be obtained through meristem culture

(A) E.A. Ball

(B) Toshio Murasighe and Folke Skoog

(C) Morel and Martin (1952)

(D) Gottlieb Haberlandt (1898)

109. Production of true to type plants achieved through____________

(A) Protoplast culture

(B) Anther culture

(C) Micropropagation

(D) Meristem tip culture

110. Somatic hybridization achieved through ____________

(A) Protoplast culture

(B) Anther culture

(C) Micropropagation

(D) Meristem tip culture

111. In India larger production of TC plants ____________

(A) Somaclonal variation

(B) Banana

(C) Secondary metabolite

(D) Ornamental plants

112. Different flower color achieved through ____________

(A) Somaclonal variation

(B) Banana

(C) Secondary metabolite

(D) Ornamental plants

113. In India which institute involving and conserves the largest number of plant species under open condition ____________

(A) Jawaharlal Nehru Tropical Botanic Garden and Research Institute,

(B) NBPGR, New delhi

(C) NIPGR, New delhi

(D) BSI, Coimbatore

114. In India, which institute involving the largest number of plant species conserved under in-vitro condition ____________

(A) Jawaharlal Nehru Tropical Botanic Garden and Research Institute,

(B) NBPGR, New delhi

(C) NIPGR, New delhi

(D) BSI, Coimbatore

115. Hormone responsible for Somatic embryo maturation ____________

(A) BAP (B) GA3

(C) IBA (D) ABA

116. Which year NCS-TCP amended?

(A) 2001 (B) 1980

(C) 1991 (D) 2006

117. Which government agency authorising the NCS-TCP?

(A) ICAR (B) TNAU

(C) CSIR (D) DBT

118. Which year 'Seed Act "amended?

(A) 2001 (B) 1966

(C) 1991 (D) 2006

119. How many Referral laboratory (ies) for virus diagnosis/genetic fidelity testing of tissue cultures plants in India?

(A) 20 (B) 2

(C) 19 (D) 10

120. What are the macronutrients used in plant cell culture medium?

(A) N, P, K, S, Na

(B) N, P, K, Ca, Cl

(C) N, P, K, S, Ca

(D) N, P, Ca, Na, Cl

121. Concentration of sucrose generally used in plant cell culture media is

(A) 10-15 g/l (B) 20-30 g/l
(C) 40-50 g/l (D) 60-70 g/l

122. Which one of the method is not a cryopreservatation

(A) Over soild Carbon di oxide (-79°C)
(B) Vapour phase of Nitrogen (-150°C)
(C) Liquid Nitrogen (-196°C)
(D) Over Glycerol stock (4°C)

123. Which one of the following is not a cryoprotectant

(A) Dimethyl sulfoxide
(B) Polyvinylpyrrolidone
(C) Proline
(D) Potassium chloride

124. In indirect somatic embryogenesis, Name the third stages of it's?

(A) Globular (B) Heart
(C) Torpedo (D) Cotyledonary

125. Stimulates root, callus and embryo formation is known as

(A) Cytokinin (B) Callogenesis
(C) Rhizogenesis (D) Auxin

126. Who developed synthetic medium

(A) White (B) Gautheret
(C) Skoog (D) Hanning

127. The mature embryos of crucifers was isolated by

(A) Robbins (B) Haberlandt
(C) Hanning (D) Van overbeek

128. Virus free dhalia plant was produced by

(A) Muir (B) Miller
(C) Ball (D) Morel and Martin

129. Who coined the term totipotency?

(A) Morgan TH (B) Cocking EC
(C) Haberlandt G (D) Carrel, A

130. Who proposed the Cell Differentiation theory?

(A) Steward *et al.* (1964)
(B) Halperin (1970)
(C) Street (1976)
(D) Tisserat *et al.* (1979)

131. The secondary metabolite *Azadirachtin* used as

(A) Antimalarial
(B) Insecticidal
(C) Cardiac tonic
(D) Antibacterial, anti inflammatory

132. The secondary metabolite Berberine used as

(A) Antimalarial
(B) Insecticidal
(C) Cardiac tonic
(D) Antibacterial, anti inflammatory

133. The secondary metabolite Digoxin obtained from used as

(A) Antimalarial
(B) Insecticidal
(C) Cardiac tonic
(D) Antibacterial, anti inflammatory

134. The secondary metabolite Quinine used as

(A) Antimalarial
(B) Insecticidal
(C) Cardiac tonic
(D) Antibacterial, anti inflammatory

135. Anther culture was first obtained in which crop

(A) *Datura innoxia* (B) *Oryza sativa*
(C) *Triticum* sp (D) *Zea mays*

136. Successful induction of haploids in an important food crop is

(A) Sorghum bicholor
(B) *Oryza sativa*
(C) *Triticum* sp
(D) *Zea mays*

137. ________ can be used to bring two genetically different parents together for common desirable characters

(A) Hybridization
(B) Protoplast fusion
(C) Polymerase chain reaction
(D) Sexual reproduction

138. Which of the following agents is achieved by Protoplast fusion?

(A) Polyethylene glycol
(B) Glycerol
(C) Ethylene
(D) Nitrogen

139. The product of protoplasm fusion is a _______

(A) Uninucleated cell
(B) Multinucleated cell
(C) Homokaryon
(D) Heterokaryon

140. A plant cell without cell wall is called _______

(A) Tropoplast (B) Protoplast
(C) Chloroplast (D) Chromoplast

141. Enucleated protoplast is called _______

(A) Cytoplast (B) Cybrid
(C) Tonoplas (D) Duplast

142. The collection of experimental methods of growing large number of isolated cells under sterile condition is called _______

(A) Plant tissue culture
(B) Taxonomy
(C) Cytology
(D) Physiology

143. ________ is necessary for drying the washed glass goods

(A) Vacuum pump
(B) Hot air oven
(C) Heater
(D) Autoclave

144. Which one of the following vitamins is used in PTC medium?

(A) Pyridoxine (B) Vitamin A
(C) Vitamin C (D) Biotin

145. The in vitro culture and maintenance of an excised organ primordial or whole or part of an organ is called as _______ culture

(A) Anther (B) Callus
(C) Organ (D) Suspension

146. _______ reported first the culture of excised root tips from the aseptically germinated wheat seedlings

(A) Kotte and Robbins
(B) White and LaRue
(C) Kotte and LaRue
(D) LaRue and Robbins

147. Who reported the successful culture of root segments of aseptically germinated tomato seedlings?

(A) P.R. White (B) C.D. LaRue
(C) S.W. Loo (D) J.P. Nitsch

148. ______ reported first in vitro culture of excised flower buds

(A) P.R. White (B) C.D. LaRue
(C) S.W. Loo (D) J.P. Nitsch

149. _______ reported the culture of 5mm shoot tips of Asparagus seedlings on a medium

(A) P.R. White (B) C.D. LaRue
(C) S.W. Loo (D) J.P. Nitsch

150. ________ cultures the ovaries of Lycopersicon esculentum

(A) P.R. White (B) C.D. LaRue
(C) S.W. Loo (D) J.P. Nitsch

151. Which one of the following culture is particularly valuable in studies of the interdependence of organs for GH and other growth factors?

(A) Anther (B) Callus
(C) Organ (D) Suspension

152. __________ is the culture of excised radicle tips of aseptically germinated seeds?

(A) Anther (B) Root
(C) Organ (D) Suspension

153. _____ culture is the in vitro culture of a generally shiny special dome-like structure

(A) Anther (B) Root
(C) Organ (D) Meristem

154. High cytokinin and low auxin are used in combination for the culture of ________

(A) Shoot (B) Root
(C) Nodule (D) Organ

155. _____ is the most effective cytokinin commonly used in shoot tip or meristem culture

(A) NAA (B) 2,4-D
(C) BAP (D) Zeatin

156. _____ is the most effective auxin commonly used in shoot tip or meristem culture

(A) NAA (B) 2, 4-D
(C) BAP (D) Zeatin

157. Coconut milk and _______ are also effective for the growth of shoot apices

(A) Gibberellic acid (B) Auxin
(C) Cytokinin (D) Ethylene

158. Asexual or vegetative propagation of whole plants using tissue culture techniques is called as ________

(A) Micropropagation
(B) Organogenesis
(C) Callus culture
(D) Suspension culture

159. Higher plant body is ________

(A) Unicellular (B) Multicellular
(C) Enucleated (D) Binucleated

160. Unorganized proliferative mass of cells is called as _______

(A) Callus (B) Explant
(C) Fragments (D) Totipotent

161. Small excised portion of the _____ is used to produce mass of cells

(A) Callus (B) Explant
(C) Fragments (D) Totipotent

162. The excised plant tissues loose its _______ integrity in culture

(A) Chemical (B) Physical
(C) Structural (D) Biological

163. Who first succeeded in promoting the development of callus tissue?

(A) Gautheret (B) Nobecourt
(C) Can Overbeek (D) Conklin

164. To raising of plants from a small tissue in culture is known as

(A) Macroproduction
(B) Micropropagation
(C) Tissue culture
(D) Mass production

165. For callus culture the manipulation sequence forceps, scalpels must kept in _____% ethanol

(A) 90 (B) 95
(C) 80 (D) 85

166. The callus tissue formation is processed through _______ of the explant

(A) Cell alteration
(B) Cell lengthening
(C) Cell expansion
(D) Cell stringent

167. 2, 4-D alone is sufficient for ______ culture

(A) Callus (B) Organ
(C) Anther (D) Pollen

168. _______ is required for growth of cell

(A) Gibberrellin (B) Auxin
(C) Cytokinin (D) Ethylene

169. _____ is required for cell division

(A) Gibberrellin (B) Auxin
(C) Cytokinin (D) Ethylene

170. Callus is yellow due to synthesis of ______ pigments

(A) Carotenoid (B) Anthocyanin
(C) Chlorophyll (D) Phycocyanin

171. Callus is purple due to accumulation of ______

(A) Carotenoid (B) Anthocyanin
(C) Chlorophyll (D) Phycocyanin

172. Which one of the following is a type of culture in which single cells or small aggregates of cells multiply?

(A) Micropropagation
(B) Organogenesis
(C) Callus culture
(D) Suspension culture

173. The developmental pathway of numerous well organized small embryoids is called as ______

(A) Organogenesis
(B) Micropropagation
(C) Somatic embryogenesis
(D) Suspension culture

174. The zygote embryo enclosed by protective coat developed from integument is ______

(A) Artificial seeds (B) Embryo
(C) Anther (D) Pollen

175. Virus free plants can be obtained through

(A) Anitibiotic treatment
(B) Bordeaux micture
(C) Root tip culture
(D) Shoot tip culture

176. Artificial seeds help to study the role of ______

(A) Endocarp (B) Endosperm
(C) Pericarp (D) Mesocarp

177. Embryo is used as explant for the ______ culture

(A) Embryo (B) Anther
(C) Pollen (D) Callus

178. Which type of culture technique is used for the development of anthers?

(A) Anther (B) Pollen
(C) Callus (D) Embryo

179. Pollen culture is otherwise known as ______ culture

(A) Microspore (B) Macrospore
(C) Angiospore (D) Apospore

180. Which one of the following is used for the in vitro development of haploid plants?

(A) Organogenesis
(B) Androgenesis
(C) Embryogenesis
(D) Micropropagation

181. Microspore behaves like a zygote and undergoes change to form embryoid is known as ______

(A) Direct androgenesis
(B) Indirect androgenesis
(C) Anther culture
(D) Pollen culture

182. Which one of the following is the ideal material for anther culture?

(A) Tomato (B) Potato
(C) Tobacco (D) Soyabean

183. Nurse culture technique is the method come under ______ culture

(A) Pollen (B) Anther
(C) Callus (D) Suspension

184. _____ is a part of plant cell which lies within the cell wall and can be plasmolyed

(A) Chloroplast
(B) Tonoplast
(C) Cyanoplast
(D) Protoplast

185. Who isolated protoplast mechnically?

(A) Klercker (B) Kuster
(C) Cocking (D) Power

186. Variations observed during tissue culture of some plants are known as

(A) Clonal variations
(B) Somatic variations
(C) Somaclonal variations
(D) Plant variations

187. Which ones produce androgenic haploids in anther cultures?

(A) Anther wall
(B) Tapetal layer of anther wall
(C) Connective tissue
(D) Young pollengrains

188. The process in which two or more adjoining somatic protoplasts fused without any inducer is called as _______ fusion

(A) Spontaneous
(B) Induced
(C) Protoplast
(D) Somatic cell

189. Chemosfusion is done using _______

(A) Sodium chloride
(B) Mannitol
(C) Polyethylene glycol
(D) Agarose

190. ______ fusion is a physical phenomenon

(A) Spontaneous
(B) Induced
(C) Protoplast
(D) Somatic cell

191. Phenotypic changes are associated with ______ changes of an organism

(A) Genetic
(B) Physical
(C) Chemical
(D) Biological

192. The enzyme that can release the individual cells from the leaf tissues is ______

(A) Macerozyme
(B) Lysozyme
(C) Zymogen
(D) Chymotrypsin

193. _______ is the genetic variability regenerated during tissue culture

(A) Somoclonal variation
(B) Organogenesis
(C) Micropropagation
(D) Androgenesis

194. Who analysed somaclonal variation in a large number of plants?

(A) Larkin & Scowcraft
(B) Evans & Sharp
(C) Hein & Mee
(D) Stepar & Bidney

195. Which one of the following auxins inhibits the production of certain secondary metabolites?

(A) IAA (B) 2, 4-D
(C) NAA (D) Kinetin

196. Callus formation is more at a temperature range of ______degree celsius

(A) 22-26 (B) 22-28
(C) 22-27 (D) 22-26

197. Plant growth regulators influence growth, metabolism and ________ of cultured cells

(A) Differentiation
(B) Elongation
(C) Enlargement
(D) Duplication

198. The optimal pH of most of the tissue cultures is in the range of

(A) 2-3 (B) 3-4
(C) 4-5 (D) 5-6

199. The ability of the callus cells to differentiate into a plant organ is called as ______

(A) Dedifferentiation
(B) Differentiation
(C) Redifferentiation
(D) Totipotency

200. Who is regarded as the Father of PTC?

(A) Gottlieb Haberlandt
(B) Watson
(C) Barbara (D) Crick

ANSWER SHEET

1. **B**	2. **D**	3. **D**	4. **D**	5. **D**	6. **C**	7. **B**	8. **B**	9. **C**	10. **C**
11. **D**	12. **A**	13. **B**	14. **A**	15. **A**	16. **B**	17. **B**	18. **C**	19. **A**	20. **A**
21. **A**	22. **A**	23. **A**	24. **C**	25. **D**	26. **B**	27. **A**	28. **C**	29. **B**	30. **B**
31. **A**	32. **C**	33. **A**	34. **C**	35. **A**	36. **A**	37. **B**	38. **A**	39. **A**	40. **C**
41. **B**	42. **A**	43. **B**	44. **C**	45. **C**	46. **B**	47. **A**	48. **C**	49. **C**	50. **A**
51. **D**	52. **C**	53. **A**	54. **C**	55. **A**	56. **B**	57. **A**	58. **A**	59. **C**	60. **D**
61. **B**	62. **B**	63. **C**	64. **B**	65. **C**	66. **B**	67. **B**	68. **A**	69. **B**	70. **B**
71. **D**	72. **C**	73. **B**	74. **B**	75. **D**	76. **D**	77. **C**	78. **A**	79. **B**	80. **A**
81. **A**	82. **A**	83. **A**	84. **A**	85. **A**	86. **C**	87. **A**	88. **A**	89. **A**	90. **B**
91. **B**	92. **A**	93. **D**	94. **D**	95. **B**	96. **B**	97. **A**	98. **D**	99. **D**	100. **B**
101. **D**	102. **D**	103. **A**	104. **D**	105. **B**	106. **A**	107. **A**	108. **C**	109. **C**	110. **A**
111. **B**	112. **A**	113. **A**	114. **B**	115. **D**	116. **D**	117. **D**	118. **B**	119. **B**	120. **C**
121. **B**	122. **D**	123. **D**	124. **C**	125. **D**	126. **A**	127. **C**	128. **D**	129. **A**	130. **B**
131. **B**	132. **D**	133. **C**	134. **A**	135. **A**	136. **C**	137. **B**	138. **A**	139. **D**	140. **B**
141. **C**	142. **A**	143. **B**	144. **A**	145. **C**	146. **A**	147. **A**	148. **B**	149. **C**	150. **D**
151. **C**	152. **B**	153. **D**	154. **A**	155. **C**	156. **A**	157. **A**	158. **A**	159. **B**	160. **A**
161. **B**	162. **C**	163. **A**	164. **B**	165. **B**	166. **C**	167. **A**	168. **B**	169. **C**	170. **A**
171. **B**	172. **D**	173. **C**	174. **A**	175. **D**	176. **B**	177. **A**	178. **A**	179. **A**	180. **B**
181. **A**	182. **C**	183. **A**	184. **D**	185. **A**	186. **C**	187. **D**	188. **A**	189. **D**	190. **C**
191. **A**	192. **D**	193. **A**	194. **B**	195. **A**	196. **B**	197. **A**	198. **D**	199. **C**	200. **A**

Chapter 6

IMMUNOLOGY

1. The most scientific method to identify which triggers may be causing my nasal allergies is to:
   (A) Track the time of year when my symptoms occur
   (B) Have an allergist perform a skin test
   (C) Have an allergist run a blood test
   (D) There is no good way to know what I'm allergic to

2. Which is the most effective treatment for nasal allergies?
   (A) Over-the-counter antihistamines
   (B) Prescription antihistamines
   (C) Antihistamine decongestant combination pills
   (D) Nasal corticosteroid sprays

3. Which one of the following is least likely to trigger asthma?
   (A) Strenuous exercise
   (B) A common cold
   (C) Reading the newspaper
   (D) Cat dander

4. Activation of B cell receptor by the binding of an epitope result in the formation of
   (A) plasma cells and T cytotoxic cells
   (B) memory cells and T cytotoxic cells
   (C) Plasma cells for antibody production and memory cells for primary response
   (D) Plasma cells for antibody production and memory cells for secondary response

5. Hypersensitivity of which system of your body creates an allergic response?
   (A) Skin (B) Immune System
   (C) Circulatory System
   (D) Pulmonary System

6. Which of the following treatments can be used to reduce allergic reactions to unavoidable substances?
   (A) Milk thistle three times daily
   (B) Mist inhalation therapy
   (C) Regularly wearing a face mask
   (D) Desensitization (Allergy shots)

7. Which of the following is the most commonly prescribed treatment for hives?
   (A) Oral antihistamines
   (B) Oral corticosteroids such as prednisone
   (C) Epinephrine (adrenaline)
   (D) Aspirin

8. Which of the following does NOT reduce or delay allergies in children?
   (A) Reducing dust-collecting items
   (B) Limiting exposure early in life to indoor furry pets
   (C) Avoiding smoking in and around the house
   (D) Exclusively wearing cotton clothes

9. Which of these body systems causes allergic reactions?
   (A) Lymph (B) Immune
   (C) Nervous (D) Autonomic

10. An allergen is anything that triggers an allergic or hypersensitive response.Which of these could be an allergen?

(A) Dust (B) Food
(C) Nickel jewelry (D) Pollution

11. Primary lymphoid organs include

(A) Thymus and spleen
(B) Thymus and bone marrow
(C) Thymus, bone marrow and spleen
(D) Thymus, bone marrow, spleen and lymph nodes

12. The inappropriate response of immune system towards a relatively harmless antigen causing harm to the host is referred as

(A) Hypersensitivity
(B) Auto-immune diseases
(C) Immunodeficiency
(D) Tolerance

13. Which of the following Ig is involved in mediating allergic reactions

(A) IgG (B) IgM
(C) IgE (D) IgA

14. The major chemical messenger involved in hypersensitivity is

(A) Interleukines (B) Lymphokines
(C) Hiatamines (D) Interferons

15. Which of the following immune mechanism is responsible for protecting us from diseases of otherspecies?

(A) Active immunity
(B) Passive immunity
(C) Innate immunity
(D) Adaptive immunity

16. Which one of the following is a cell mediated hypersensitive reaction

(A) Type I (B) Type II
(C) Type III (D) Type IV

17. The inability to distinguish between self-cells and non-self-cells may lead to

(A) Hypersensitivity
(B) Auto-immune diseases
(C) Immunodeficiency
(D) Tolerance

18. Majority of auto immune diseases are

(A) Cell mediated
(B) Antibody mediated
(C) Macrophage mediated
(D) Mast cells mediated

19. All of the following are autoimmune disorders except

(A) Graves disease
(B) SCID
(C) Rheumatoid arthritis
(D) Addison's disease

20. Rheumatoid arthritis mostly occur in individuals carrying

(A) HLA-DR4 gene (HLA-human leucocyte antigen)
(B) HLA-DR1 gene
(C) HLA-DR3 gene
(D) HLA-DR2 gene

21. Some defects or mutations in components of innate or adaptive immunity may lead to

(A) hypersensitivity
(B) auto-immune diseases
(C) immunodeficiency
(D) tolerance

22. In severe combined immune deficiency (SCID), the patients are deficient in

(A) B cells (B) T cells
(C) Both A & B (D) IgA

23. SCID can occur due to the absence of an enzyme

(A) Adenosine deaminase
(B) Guanosine deaminase
(C) Phosphorylase
(D) Thymidine deaminase

24. HIV attacks

(A) T helper cells
(B) T cytotoxic cells
(C) B cells
(D) Macrophages

25. All of the following are immunodeficiency diseases except

(A) Graves disease
(B) SCID
(C) DiGeorge's syndrome
(D) Hyper IgM syndrome

26. Injection of anti-venom to a patient for snake bite is an example of

(A) Naturally acquired active immunity
(B) Artificially acquired active immunity
(C) Naturally acquired passive immunity
(D) Artificially acquired passive immunity

27. Newborns get their antibodies from mothers milk. This is an example of

(A) Naturally acquired active immunity
(B) Artificially acquired active immunity
(C) Naturally acquired passive immunity
(D) Artificially acquired passive immunity

28. Vaccination is an example of

(A) Naturally acquired active immunity
(B) Artificially acquired active immunity
(C) Naturally acquired passive immunity
(D) Artificially acquired passive immunity

29. Often patients are immune to diseases like chicken pox once infected. This immunity is an example of

(A) Naturally acquired active immunity
(B) Artificially acquired active immunity
(C) Naturally acquired passive immunity
(D) Artificially acquired passive immunity

30. Skin is the largest organ in the body and protects us by preventing pathogen entry. This is an example of

(A) Naturally acquired innate immunity
(B) Artificially acquired adaptive immunity
(C) Naturally acquired passive immunity
(D) Artificially acquired passive immunity

31. All the following comes under non-specific defence mechanism except

(A) Fever
(B) Phagocytes
(C) Cell mediated immunity
(D) Complement system

32. Which of the following is the site of T cell maturation?

(A) Bone marrow (B) Thymus
(C) Spleen (D) Appendix

ANSWER SHEET

1. **B/C**	2. **D**	3. **B**	4. **D**	5. **C**	6. **D**	7. **A**	8. **D**	9. **B**	10. **B**
11. **B**	12. **A**	13. **C**	14. **C**	15. **C**	16. **D**	17. **B**	18. **B**	19. **B**	20. **A**
21. **C**	22. **C**	23. **A**	24. **A**	25. **A**	26. **D**	27. **C**	28. **B**	29. **A**	30. **A**
31. **C**	32. **B**								

Chapter - 7

ANIMAL BIOTECHNOLOGY

1. In animal cell cultures, the addition of serum to media is essential for providing?
   (A) Growth factors
   (B) Amino acid for protein synthesis
   (C) Nucleotide for DNA synthesis
   (D) Nutrients
2. CDC mutant useful for the study of
   (A) Various stages of cell cycle
   (B) Homeodomain
   (C) Apoptosis only
   (D) Chromosome break point
3. Recombinant DNA human health care product eveloped for treating haemophiliacs is
   (A) Interferon (B) Lymphokines
   (C) Coaglutation factor IX
   (D) Phytohemaglutinin
4. Which of the following has been produced commercially from mammalian culture?
   (A) Insulin (B) Renin
   (C) Plasminogen activator
   (D) Antibacterial antibody
5. A heterologous protein for its expression in the milk of a transgenic animal should be under the control of the promoter of the gene coding for
   (A) Preproinsulin (B) Lac Z
   (C) β globin (D) β lactoglobulin
6. Antibody diversity is generated by
   (A) Allelic exclusion
   (B) Protein splicing
   (C) Somatic mutation
   (D) Interchromosomal recombination
7. Reverse vaccinology indicates
   (A) From antibody to vaccine development
   (B) From genome sequence to vaccine development
   (C) From antigenic polysaccharide to vaccine development
   (D) From antigenic protein to vaccine development
8. The culture fluid of 1000 to 5000 colonies of hybridoma are screened for monoclonal antibody by
   P- Antigen capture analysis Q- Western blot analysis
   R-Northern blot analysis S-antibody capture analysis
   (A) P, Q (B) R, S
   (C) Q, S (D) P, S
9. All of the following are produced by animal cells in culture and help the cells adhere to the culture dish except
   (A) Collagen (B) Glycoprotein
   (C) Phospholipase A
   (D) Hyaluronic acid

ANSWER SHEET

1. A 2. A 3. C 4. C 5. D 6. A 7. D 8. D 9. C

Chapter 8

BIOINFORMATICS

1. The term bioinformatics was coined by
   (A) J D Watson
   (B) Margaret Dayhoff
   (C) Pauline Hogeweg
   (D) Frederic Sanger

2. Margaret Dayhoff developed the first protein sequence database called
   (A) SWISS PROT
   (B) PDB
   (C) Atlas of protein sequence and structure
   (D) Protein sequence databank

3. Step wise method for solving problems in computer science is called
   (A) Flowchart (B) Sequential design
   (C) Procedure (D) Algorithm

4. The first published completed gene sequence was of
   (A) M 13 phage
   (B) T 4 phage
   (C) φ X174
   (D) Lambda phage

5. The term used to refer something 'performed on computer or computer simulation"
   (A) Dry lab (B) Web lab
   (C) In vitro (D) Insilico

6. 'Laboratory work using chemicals, drugs etc using water' is referred as
   (A) Dry lab (B) Web lab
   (C) Wet lab (D) Insilico

7. 'Laboratory work using computers and computer generated models generally offline' is referred as
   (A) Dry lab
   (B) Web lab
   (C) Wet lab
   (D) Insilico

8. 'Laboratory work using computers and associated web based analysis generally online' is referred as
   (A) Dry lab (B) Web lab
   (C) Wet lab (D) Insilico

9. 'Invitro' in latin means
   (A) Within the glass
   (B) Within the lab
   (C) Outside the lab
   (D) Outside the glass

10. NCBI was established in
    (A) 1988 (B) 1989
    (C) 1990 (D) 1991

11. Proteomics research can be categorized as
    (A) Structural proteomics and functional proteomics
    (B) Structural, functional and comparative proteomics
    (C) Functional and comparative proteomics
    (D) Structural proteomics

12. The computational methodology that tries to find the best matching between two molecule, a receptor and ligand is called

(A) Molecular matching
(B) Molecular docking
(C) Molecular fitting
(D) Molecule affinity checking

13. Proteomics is the study of

(A) Set of proteins
(B) Set of proteins in a specific region of the cell
(C) Entire set of expressed proteins in a cell
(D) Set of proteins with specific function

14. The process of finding relative location of genes on a chromosome is called

(A) Gene tracing
(B) Genome mapping
(C) Genome walking
(D) Chromosome walking

15. A compound that has desirable properties to become a drug is called

(A) Lead
(B) Find
(C) Fit drug
(D) Fit compound

16. Margaret Dayhoff developed the first protein sequence database called

(A) SWISS PROT
(B) PDB
(C) Atlas of protein sequence and structure
(D) Protein sequence databank

17. Each record in a database is called an

(A) Entry (B) File
(C) Record (D) Ticket

18. Literature databases include

(A) MEDLINE and PubMED
(B) MEDLINE and PDB
(C) PubMED and PDB
(D) MEDLINE and PDS

19. *FlyBase* is a

(A) Biodiversity database
(B) Model organism database
(C) Literature database
(D) Biomolecular database

20. Which of the following is an *E. coli* model organism database

(A) EcoGene (B) EcoBase
(C) EcoSeq (D) ColGene

21. Which of the following is a protein sequence database

(A) DDBJ (B) EMBL
(C) GenBank (D) PIR

22. GenBnak, the nucleic acid sequence database is maintained by

(A) Brookhaven laboratory
(B) DNA database of Japan (DDBJ)
(C) European Molecular Biology laboratory (EMBL)
(D) National Centre for Biotechnology Information (NCBI)

23. Submission to GenBanK are made using

(A) BankIt and Sequin
(B) BankIt and BankIn
(C) Sequin and BankIn
(D) Entrez

24. STAG is maintained by

(A) Brookhaven laboratory
(B) DNA database of Japan (DDBJ)
(C) European Molecular Biology laboratory (EMBL)
(D) National Centre for Biotechnology Information (NCBI)

25. A comprehensive database for the study of human genetics and molecular biology is

(A) PDB (B) STAG
(C) OMIM (D) PSD

26. All the following are protein sequence databases except

(A) PIR (B) PSD
(C) SWISS PROT (D) EMBL

27. The information retrieval tool of NCBI GenBank is

(A) Entrez (B) STAG
(C) SeqIn (D) Text search

28. The first secondary database developed

(A) PRINTS (B) PROSITE
(C) PDB (D) PIR

29. Which of the following is a sequence alignment tool

(A) BLAST (B) PRINT
(C) PROSITE (D) PIR

30. MAtDB is a model organism database for

(A) Mouse (B) Human
(C) *E. coli* (D) Arabidopsis

31. Which is model organism database?

(A) GOLD (B) PROMISE
(C) SGD (D) SCOP

32. BLAST X program is used for

(A) Translate protein sequence
(B) Translate DNA database
(C) Translate input sequence
(D) Translate protein and nucleotide sequence

33. Which is data retrieving tool?

(A) ENTREZ (B) EMBL
(C) PHD (D) PDB

34. GeneBank and SWISSPORT are example of

(A) Primary database
(B) Secondary database
(C) Composite database
(D) Structural database

35. SCOP is

(A) It is primary database
(B) It is nucleotide sequence database
(C) SCOP database is a hierarchical classification of protein 2D domain structures
(D) Structural database, which identity strual and evolutionary relationships

36. FASTA' was published by

(A) Joseph Sambrook
(B) Pearson and Lipman
(C) Sanger (D) Altschul *et al*

37. PDB is

(A) Primary database for macromolecules
(B) Can be determined by gel electrophoresis
(C) Composite database
(D) Database for three dimensional structure of biological macromolecule

38. BLOSUM matrices are used for

(A) Phylogenetic analysis
(B) Multiple sequence alignment
(C) Pairwise sequence alignment
(D) Global alignment

ANSWER SHEET

1. **C**	2. **C**	3. **D**	4. **C**	5. **D**	6. **C**	7. **A**	8. **B**	9. **A**	10. **A**
11. **A**	12. **B**	13. **C**	14. **B**	15. **A**	16. **C**	17. **A**	18. **A**	19. **B**	20. **A**
21. **D**	22. **D**	23. **A**	24. **B**	25. **C**	26. **D**	27. **A**	28. **B**	29. **A**	30. **D**
31. **C**	32. **C**	33. **A**	34. **A**	35. **D**	36. **D**	37. **A**	38. **C**		

Part 3
SEED SCIENCE AND TECHNOLOGY

Unit 1
REPRODUCTIVE BIOLOGY AND SEED BIOLOGY

Chapter 1

GYMNOSPERM

1. In which of the following group would you place a plant which produce seeds but lacks flower
   (A) Fungi (B) Pteridophytes
   (C) Bryophytes (D) Gymnosperms
2. A gymnospermic plant
   (A) Bear flower
   (B) Exhibit no vascular tissue
   (C) Produce seeds in cones
   (D) Does not produce seeds in cones
3. Megasporophyll is the term used in gymnosperm to denote
   (A) Carpels (B) Stamens
   (C) Leaves (D) Female cone
4. The megaspangrium is also known as
   (A) Ovule (B) Nucellus
   (C) Fruit (D) Micropyle
5. Perisperm is
   (A) Outgrowth of the outer integument
   (B) Surviving nucellus in the seed
   (C) Outgrowth of funicles
   (D) Outgrowth of micropyle
6. Gymnosperm differ from angiosperm
   (A) Having seeds (B) Having fruits
   (C) Having naked ovules
   (D) No embryo
7. Alternation of generation is exhibited by
   (A) Bryophytes (B) Pteridophytes
   (C) Gymnosperms (D) All plants
8. In which of the following feature angiosperm resemble gymnosperm
   (A) Presence of ovule
   (B) Presence of vessel
   (C) Nature of endosperm
   (D) Mode of fertilization
9. Endosperm in gymnosperm is formed
   (A) At the time of fertilization
   (B) Before fertilization
   (C) After fertilization
   (D) Along with the development of embryo
10. In gymnosperm pollination if exclusively by
    (A) Animals (B) Wind
    (C) Insects (D) Water
11. Largest sperms are found in
    (A) Pinus (B) Cedrus
    (C) Cycas (D) Gnetum
12. Zooidogamy is seen in
    (A) Cycas (B) Gnetum
    (C) Pinus (D) Angiosperm

ANSWER SHEET

1. **D** 2. **C** 3. **A** 4. **A** 5. **B** 6. **C** 7. **D** 8. **B** 9. **B** 10. **B** 11. **C** 12. **A**

Chapter 2

ANGIOSPERM

1. An inflorescence in which flowers arise from different point but reach at same point is known as

   (A) Catkin (B) Spadix
   (C) Umbel (D) Cymose

2. In raceme, the flowers are

   (A) Arranged in basipetal succession
   (B) Arranged in acropetal succession
   (C) Of separate sexes
   (D) Of the same sexes

3. A spike of unisexual flowers is found in

   (A) Wheat (B) Onion
   (C) Mulberry (D) Fig

4. In onion, the inflorescence is

   (A) Raceme (B) Corymb
   (C) Catkin (D) Umbel

5. In musaceae, the inflorescence is

   (A) Spadix (B) Spike
   (C) Head (D) Umbel

6. The inflorescence in *Mimosa pudica* (touch me not plant) is

   (A) Umbel (B) Corymb
   (C) Spike (D) Globose head

7. Hypanthodium is a characteristic inflorescence of

   (A) Mulberry (B) Ficus
   (C) Poinsetia (D) Pine apple

8. The inflorescence in *Dorstenia*

   (A) Amentum (B) Coenanthium
   (C) Anthodium (D) Rhiphodium

9. When dichasial cyme ends into monochasial cyme, the inflorescence is

   (A) Uniparous (B) Biparous
   (C) Cincissus (D) Veticillaster

10. Dichasial cyme is seen in

    (A) *Ixora* (B) *Begonia*
    (C) *Calotropis* (D) *Bignonia*

11. Cyathium is found in

    (A) *Ficus* (B) *Morus*
    (C) *Euphorbia* (D) *Dorstenia*

12. Helicoid cyme is found in

    (A) *Dianthus* (B) *Hamelia*
    (C) *Leucas* (D) *Hibiscus*

13. The characteristic inflorescence of the family compositae

    (A) Head or Capitulam
    (B) Cymose
    (C) Racemose
    (D) Verticillaster

14. The inflorescence of Anthurium

    (A) Spike
    (B) Racemose
    (C) Hypanthodium
    (D) Spadix

15. The most important function of an inflorescence is to help in

(A) Dispersal of seeds
(B) Release of pollen grains
(C) Attracting insects for cross pollination
(D) Forming large number of fruits

16. A most attracting characteristic of angiosperm is

(A) Root (B) Seed
(C) Flower (D) Stem

17. A flower with essential and non-essential whorls.

(A) Incomplete (B) Irregular
(C) Sessile (D) Complete

18. A bisexual flowers which never open in its life span is called

(A) Homogamous (B) Heterogamous
(C) Polygamous (D) Cleistogamous

19. Which of the following is not a single flower?

(A) China rose (B) Petunia
(C) Sunflower (D) Holy hope

20. Plants which flower only once in their life is

(A) Polycarpic (B) Monocarpic
(C) Cleistocarpic (D) Pericarpic

21. Third whorl in flower is of

(A) Petal (B) Sepal
(C) Stamen (D) Pistil

22. Flower is a modified shoot because

(A) In some flowers thalamus become elongated and shows distinct nodes and internodes
(B) Flowers aggregate to form inflorescence
(C) Epicalyx is present in some flowers
(D) Essential organs are present in some flowers

23. In monoecious plants

(A) Male and female parts are borne by the same plant but not by the same flower
(B) Male and female parts are borne by the same flower
(C) Male and female parts are borne by the different plant
(D) None of the above

24. A flower is said to be complete when it has

(A) Corolla
(B) Androecium and Gynoecium
(C) Calyx and Corolla only
(D) All the four whorls

25. The perianth is the term used when

(A) Androecium and gynoecium are similar
(B) Calyx and corolla are similar
(C) Androecium and calyx are similar
(D) Gynoecium and corolla are similar

26. A flower is brightly coloured, scented and secrete nectar. It is most probably

(A) Pollinated by insects
(B) Sterile
(C) Pollinated by wind
(D) An insectivorous plant

27. Protandry is the situation when

(A) Anthers and stigma mature at the same time
(B) Anthers mature later than the stigma of the same flower
(C) Anthers mature earlier than the stigma of the same flower
(D) Anthers of the flower pollinate the stigma of the same flower

28. In a bisexual flowers when the gynoecium matures earlier than the androecium it is called

(A) Protogyny (B) Protandry
(C) Autogamy (D) Heterogamy

29. In which of the following plants do the male and female flowers occur in the same individual

(A) Pumpkin (B) Gram
(C) Rose (D) Hibiscus

30. In a flower, when the ovary is situated on the torus above all other floral whorls, the flower is said to be

(A) Perignyous (B) Epigynous
(C) Inferior (D) Hypogynous

31. Which of the following constituents the best definition of a fruit? A fruit is

(A) A product of the flower
(B) A product of ovary
(C) Post fertilization product of pistil
(D) A body that contains seed

32. A fruit which develops from a condensed inflorescence is called

(A) A simple fruit
(B) An eaterio of fruit
(C) An aggregate fruit
(D) A composite fruit

33. Multiple fruits are those which are produced by

(A) Several ripened ovaries of a single flower
(B) Several ripened ovaries of several flowers of an inflorescence
(C) Part other than ovary
(D) Fusion of calyx with ovary

34. Drupe is recognized by

(A) Fleshy seed coat
(B) Stony endocarp
(C) Thin seed coat
(D) Stony mesocarp

35. The edible portion of mango is

(A) Endocarp (B) Mesocarp
(C) Epicarp (D) Pericarp

36. The edible portion of coconut is

(A) Endosperm (B) Mesocarp
(C) Seed coat (D) Pericarp

37. A simple one seeded fruit with pericarp fused with the seed coat is known as

(A) Caryopsis (B) Achene
(C) Nut (D) Cypsela

38. In pomegranate the edible part is

(A) Juice epicarp (B) Mesocarp
(C) Juice hairs (D) Juice testa

39. Aggregate fruit is defined as the one which develops from

(A) Multicarpellary apocarpus gynoecium
(B) Multicarpellary syncarpous gynoecium
(C) Multicarpellary superios ovary
(D) Complete inflorescence

40. Lotus fruit is

(A) Capsule (B) Berry
(C) Etaerio of berries
(D) Etaerio of achenes

41. Betelnut is

(A) Drupe (B) Berry
(C) Nut (D) Caryopsis

42. A fruit developed from a spikate inflorescence is called

(A) Syconium (B) Sorosis
(C) Elaterio of achenes
(D) Elaterio of berries

43. A fruit which is monocarpellary, superior and dehisces by one suture only is

(A) Legume (B) Regma
(C) Follicle (D) Siliqua

44. A fruit which develops from a condensed inflorescence is called

(A) A simple fruit
(B) An etaerio of fruit
(C) An aggregate fruit
(D) A composite fruit

45. The fruit of gramineae is
(A) Caruncle (B) Caryopsis
(C) Achene (D) Nut

46. Double fertilization is characteristic of
(A) Gymnosperms
(B) Angiosperms
(C) Monocots
(D) Bryophytes

47. In angiosperm, endosperm is
(A) Haploid (B) Diploid
(C) Triploid (D) Tetraploid

48. Porogamy is
(A) Fertilization in which pollen tube enters the ovule through integument
(B) Fertilization without pollen grain
(C) Fertilization in which pollen tube enters the ovule through chalaza
(D) Fertilization in which pollen tube enters the ovule through micropyle

49. Monosporic eight nucleated female gametophyte is found in
(A) Adoxa (B) Onion
(C) Fritillaria (D) Polygonum

50. In angiosperm, the free nuclear divisions take place during
(A) Gamete formation
(B) Flower formation
(C) Endosperm formation
(D) Embryo formation

51. The name perisperm is given to the
(A) Remnant of the nucellus
(B) Peripheral nucellus
(C) Distintegrated antipodals
(D) Disintegrated synergids

52. An embryo may sometimes develop from a cell of an embryo sac other than the egg. This is called
(A) Apospory (B) Parthenogenesis
(C) Parthenocarpy (D) Apogamy

53. Double fertilization was discovered by
(A) Hofmeister (B) Leenuwenhoeck
(C) Nawaschin (D) Hooke

54. After fertilization, the seed coat develop from
(A) Chalaza (B) Ovule
(C) Embryosac (D) Integuments

55. Amphimixis in plants means development not a plant
(A) From fusion of two gametes
(B) Without fusion of gametes
(C) From stem cuttings
(D) From root cuttings

56. One of the following shows one celled suspensor
(A) Wheat (B) Petunia
(C) Hedera (D) Solanum

57. Which of the following statement is true for the pollen tube?
(A) It shows only tip growth
(B) It is composed of three non cellular zones
(C) It shows chemostatic movements
(D) It shows radial cytoplasmic streaming

58. In the anther wall
(A) Tapetum lies just inner to the endothecium
(B) Endothecium lies inner to the middle layers
(C) Tapetum lies next to the epidermis
(D) Middle layers lie between endothecium and tapeum

59. Fertilization is the process of
(A) Transfer the pollen from anther to stigma
(B) Fusion of one male gamete with the egg
(C) Formation of seed from ovule
(D) Fusion of male nucleus with polar nuclei

60. In flowering plants, meiosis occurs at the time of

(A) Germination of seed
(B) Formation of buds
(C) Formation of root primordia
(D) Formation of pollen grains

61. In an angiosperm, male plant is diploid and female plant tetraploid, endosperm will be

(A) Haploid (B) Triploid
(C) Tetraploid (D) Pentaploid

62. The arrangement of the nuclei in a normal embryo sac in the plants is

(A) 2+3+3 (B) 3+2+3
(C) 2+4+2 (D) 3+3+2

63. Apomictic embryos in citrus arise from

(A) Maternal sporophytic tissue in ovule
(B) Synergids
(C) Diploid egg
(D) Antipodal cells

64. Fibrous thickenings of hygroscopic nature are found in this part of the anther wall

(A) Middle layers
(B) Endothecium
(C) Epidermis
(D) Tapetum

65. In a type of apomixes known as adventives embryony, embryos develop directly from the

(A) Zygote
(B) Nucellus or integuments
(C) Synergids or antipodals in an embryo sac
(D) Accessory embryo sacs in the ovule

66. In nature, cleistogamous flowers are

(A) Wind pollinated
(B) Insect pollinated
(C) Self pollinated
(D) Bird pollinated

67. Which one of the following is surrounded by a callose wall?

(A) Male gamete
(B) Microspore mother cell
(C) Pollen grain
(D) Egg

68. Study the following table and select the most correct option

List I (Structure before seed)	List II (Structure after seed)
(A) Funiculus	i) Hilum
(B) Scar of ovule	ii) Tegmen
(C) Zygote	iii) Testa
(D) Inner integument	iv) Stalk of seed
	v) Embryo

(A) A-V, B-I, C-II, D-IV
(B) A-III, B-IV, C-I, D-V
(C) A-IV, B-I, C-V, D-II
(D) A-IV, B-V, C-III, D-II

69. A completely straight ovule with the micropyle at the apex is known as

(A) Amphitropous ovule
(B) Orthotropous ovule
(C) Anatropous ovule
(D) Campylotropous ovule

70. What is the function of germ pore?

(A) Initiation of pollen tube
(B) Absorption of water for seed germination
(C) Emergence of radicle
(D) Release of male gametes

71. Which of the following pairs in angiosperms are diploid and triploid respectively?

(A) Polar nucleus and secondary nucleus
(B) Microspore mother cell and egg cell
(C) Secondary nucleus and endosperm
(D) Endosperm and antipodal cells

72. Which of the following statements about sporopollenin is false?

(A) Exine has apertures called germ pores where sporopollenin is present

(B) Sporopollenin is one of the resistant organic materials

(C) Exine is made up of sporpollenin

(D) Sporopollenin can withstand high temperatures and strong acids

73. Which one of the following is not a correct explanation of cross pollination?

(A) The pollen grains of male flowers are transferred to the stigma of the female flowers

(B) The pollen grains are transferred from one flower to another flower, or another plant of the same species

(C) Transfer of pollen from a flower of one field to stigma of a flower in another field of same species

(D) Trasfer of pollen from anther to stigma of same flower

74. Which one of the following pairs of plant structures has haploid number of chromosomes?

(A) Egg cell and antipodal cells

(B) Megaspore mother cell and antipodal cells

(C) Egg nucleus and secondary nucleus

(D) Nucellus and antipodal cells

75. Which one of the following statements is wrong?

(A) Intine is made up of cellulose and pectin

(B) Vegetative cell is larger than generative cells

(C) Pollen grains in some plants remain viable for months

(D) When pollen is shed at two-celled stage, double fertilization does not take place

76. Assertion : 7-celled, 8 nucleate and monosporic embryo sac is called Polygonum type of embryo sac.

Reason : It was discovered by Hofmeister for the first time in Polygonum.

(A) Both Assertion and Reason are true and Reason is the correct explanation of the Assertion

(B) Both Assertion and Reason are true but the Reason is not the correct explanation of Assertion

(C) Assertion is true, but Reason is false.

(D) Both Assertion and Reason are false.

77. In the fully organized polygonum type of embryo sac, what is the ratio of haploid, diploid and triploid nuclei?

(A) 6:1:0 (B) 6:0:1

(C) 3:1:3 (D) 3:2:3

78. Vernalisation stimulates flowering in

(A) Carrot (B) Turmeric

(C) Elephant Yam (D) Ginger

79. What is the ratio of equational divisions that take place in Cycas and angiosperms respectively leading to the formation of male gametes from pollen grains?

(A) 2:1 (B) 3:1

(C) 3:2 (D) 2:3

80. Ornithophilly refers to the pollination by which of the following :

(A) Snails (B) Birds

(C) Insects (D) Air

81. Ovule is attached to the placenta of ovary wall by

(A) Funicle (B) Raphae

(C) Hilum (D) Testa

82. Plants with ovaries having only one or a few ovules, are generally pollinated by

(A) Birds (B) Butterflies

(C) Bees (D) Wind

83. Sporopollenin is chemically
(A) Protein
(B) Fatty substance
(C) Homopolysaccharide
(D) Heteropolysaccharide

84. The pollen are liberated in Cassytha by
(A) Transverse dehiscence
(B) Longitudinal dehiscence
(C) Porous dehiscence
(D) Valvular dehiscence

85. There are 10 flowers in one individual plant of Crotalaria. In each microsporangium of every stamen of all the flowers, there are 30 microspore mother cells. How many pollen grains formed from that plant?
(A) 48,000
(B) 24,000
(C) 10,000
(D) 4,000

86. Consider the following statements and choose the correct option.
(A) The genetic constitution of a plant is unaffected in vegetative propagation
(B) Rhizome in ginger serves as an organ of vegetative reproduction
(C) Totipotency of cells enables us to micropropagate plants

(A) Statements (B) alone is true
(B) Statement (A) and (B) alone are true
(C) Statements (B) and (C) alone are true
(D) All the three statements (A), (B) and (C) are true

87. Double fertilization involves :
(A) Fertilization of the egg by two male gametes
(B) Fertilization of two eggs in the same embryo sac by two sperms brought by one pollen tube.
(C) Fertilization of the egg and the central cell by two sperms brought by different pollen tubes.
(D) Fertilization of the egg and the central cell by two sperms brought by the same pollen tube

88. Grafting is successful in dicots, but not in monocots because the dicots have :
(A) Cork cambium
(B) Vessels with elements arranged end to end
(C) Vascular bundles arranged in a ring
(D) Cambium for secondary growth

89. In which one pair both the plants can be vegetatively propagated by leaf pieces?
(A) Asparagus and Bryophyllum
(B) Agave and Kalanchoe
(C) Bryophyllum and Kalanchoe
(D) Chrysanthemum and Agave

90. Match the following ovular structure with post fertilization

A. Ovule	1. Endosperm
B. Funiculus	2. Aril
C. Nucellus	3. Seed
D. Polar nuclei	4. Perisperm

(A) A=2, B=3, C=4, D=1
(B) A=2, B=3, C=1, D=4
(C) A=3, B=2, C=4, D=1
(D) A=3, B=2, C=1, D=4

91. Pick out the wrong statement.
(A) Double fertilization is unique to gymnosperms and monocotyledons
(B) Sequoia, a gymnosperm, is one of the tallest trees.
(C) Phaeophyceae members possess chlorophyll a, c, carotenoids and xanthophylis
(D) Moss is gametophyte which consists of two stages namely, protonema stage and leafy stage.

92. An organic substance that can withstand environmental extremes and can not be degraded by any enzyme is

(A) Lignin (B) Cellulose
(C) Cuticle (D) Sporopollenin

93. The coconut water and the edible part of coconut are equivalent to

(A) Mesocarp (B) Embryo
(C) Endosperm (D) Endocarp

94. The gynoecium consists of many free pistils in flowers of

(A) Papaver (B) Michelia
(C) Aloe (D) Tomato

95. Both, autogamy and geitonogamy are prevented in

(A) Castor (B) Maize
(C) Papaya (D) Cucumber

96. Even in absence of pollinating agents, seed setting is assured in

(A) Salvia (B) Fig
(C) Commelina (D) Zostera

97. Wind pollinated flowers are

(A) Small, producing large number of dry pollen grains
(B) Small, brightly coloured, producing large number of pollen grains
(C) Small, producing nectar and dry pollen
(D) Large, producing abundant nectar and pollen

98. In anatropous ovule, what is the direction of micropyle?

(A) Right (B) Downward
(C) Upward (D) Left

99. Which type of association is found in between entomophilous flower and pollinating agent?

(A) Cooperation (B) Commensalism
(C) Mutualism (D) Coevolution

100. Transfer of pollen grains from the anther to the stigma of another flower of the same plant is called

(A) Geitonogamy (B) Xenogamy
(C) Autogamy (D) Karyogamy

101. In which one of the following, male and female gametophytes do not have free living independent existence?

(A) Funaria (B) Pteris
(C) Cedrus (D) Polytrichum

102. When the other floral parts are arranged at the base of the gynoecium, the flower is called :

(A) Perigynous flower
(B) Epigynous flower
(C) Hypogynous flower
(D) None of these

103. Male gametes in angiosperms are formed by the division of :

(A) Generative cell
(B) Microspore
(C) Microspore mother cell
(D) Vegetative cell

104. The aleurone layer in maize grain is specially rich in :

(A) Lipids (B) Starch
(C) Proteins (D) Auxins

105. Endosperm is formed during the double fertilization by

(A) Ovum and male gamete
(B) One polar nuclei and one male Gamete
(C) Two polar nuclei and one male gamete
(D) Two polar nuclei and two male gametes

106. In Angiosperm, pollen tube liberate their male gametes into the

(A) Egg cell (B) Antipodal cells
(C) Central cell (D) Synergids

107. A clone is a group of individuals obtained through
(A) Self pollination
(B) Vegetative propagation
(C) Hybridization
(D) Cross pollination

108. Filiform apparatus is a characteristic feature of
(A) Zygote (B) Suspensor
(C) Egg (D) Synergid

109. A monocarpic plant is one which
(A) Has only one carpel
(B) Flowers and fruits only once in life time and thereafter dies
(C) Produces only one seed
(D) None of these

110. Axillary buds are used to raise crop of
(A) Wheat (B) Rice
(C) Groundnut (D) Sugarcane

111. Banana plant develops from
(A) Rhizome (B) Seed
(C) Sucker (D) Stolon

112. In a flowering plant, archesporium gives rise to
(A) Only the wall of the sporangium
(B) Both wall and the sporogenous cells
(C) Wall and the tapetum
(D) Only tapetum and sporogenous cells

113. In which one of the following, pollination is autogamous?
(A) Cleistogamy (B) Geitonogamy
(C) Xenogamy (D) Chasmogamy

114. Insect pollinated flowers are characterized by
(A) Large number of pollens
(B) Dry and smooth pollens
(C) Sticky and rough pollens
(D) Heavy pollens

115. 8-nucleate embryosacs are
(A) Monosporic only
(B) Bisporic only
(C) Tetrasporic only
(D) Any of these

116. A bisexual flower which never opens up in its life span is known as
(A) Autogamy (B) Allogamy
(C) Homogamy (D) Cleistogamy

117. A hyaline, bisexual and self-fertilized flower that does not open at all, is
(A) Chasmogamous (B) Apogamous
(C) Cleistogamous (D) Polygamous

118. A polygonum type of embryo sac is
(A) 7-celled and 8-nucleate
(B) 8-celled and 7-nucleate
(C) 7-celled and 7-nucleate
(D) 8-celled and 8-nucleate

119. Allogamy is best favoured by
(A) Dichogamy (B) Dicliny
(C) Cleistogamy (D) Homogamy

120. An interesting modification of flower shape for insect pollination occurs in some orchids in which a male insect mistakes the pattern on the orchid flower for the female of his species and tries to copulate with it, thereby pollinating the flower. This phenomenon is called
(A) Pseudo parthenocarpy
(B) Pseudocopulation
(C) Pseudo pollination
(D) Pseudo anthesis

121. An orthotropous ovule is one in which micropyle and chalaza are
(A) In straight line of funicle
(B) Parallel along with funicle
(C) At right angles to funicle
(D) Parallel along with ovule

122. Anemophillous flowers have

(A) Sessile stigma
(B) Small smooth stigma
(C) Coloured flower
(D) Large feathery stigma

123. Anthesis is

(A) Opening of flower bud
(B) Pollen mother cell undergoes meiosis
(C) Dehiscence of anthers
(D) Stigma becoming receptive

124. Double fertilization involves

(A) Fusion of secondary nucleus with both male gametes
(B) Fusion of secondary nucleus with one male gamete
(C) Fusion of one polar nucleus with one male gamete
(D) Any of above, depending on species

125. Cauliflory is flowers produced

(A) An old dormant bud
(B) On young bud
(C) On the axil
(D) On the branch

126. Cleistogamous condition is present in

(A) Brassica
(B) *Solanum tuberosum*
(C) Commelina
(D) *Allium cepa*

127. In Angiosperms :

(A) Sporophyte is complex and gametophyte inconspicuous
(B) Sporophyte simple and gametophyte complex
(C) Sporophyte and gametophyte complex
(D) Sporophyte and gametophyte inconspicuous

128. An embryo-sac of angiosperms generally contains :

(A) Two nuclei (B) Three nuclei
(C) Four nuclei (D) Eight nuclei

ANSWER SHEET

1. **C**	2. **B**	3. **C**	4. **D**	5. **A**	6. **B**	7. **B**	8. **B**	9. **D**	10. **A**
11. **C**	12. **B**	13. **A**	14. **D**	15. **C**	16. **C**	17. **D**	18. **D**	19. **C**	20. **B**
21. **C**	22. **A**	23. **A**	24. **D**	25. **B**	26. **A**	27. **C**	28. **A**	29. **A**	30. **D**
31. **B**	32. **D**	33. **B**	34. **B**	35. **B**	36. **A**	37. **A**	38. **D**	39. **A**	40. **D**
41. **B**	42. **B**	43. **C**	44. **D**	45. **B**	46. **B**	47. **C**	48. **D**	49. **D**	50. **C**
51. **A**	52. **D**	53. **C**	54. **D**	55. **A**	56. **A**	57. **A**	58. **D**	59. **B**	60. **D**
61. **B**	62. **B**	63. **A**	64. **B**	65. **B**	66. **C**	67. **B**	68. **C**	69. **B**	70. **A**
71. **C**	72. **A**	73. **D**	74. **A**	75. **D**	76. **C**	77. **A**	78. **A**	79. **A**	80. **B**
81. **A**	82. **D**	83. **B**	84. **D**	85. **A**	86. **D**	87. **D**	88. **D**	89. **C**	90. **C**
91. **A**	92. **D**	93. **C**	94. **B**	95. **C**	96. **C**	97. **A**	98. **B**	99. **C**	100. **A**
101. **A**	102. **C**	103. **A**	104. **C**	105. **C**	106. **D**	107. **B**	108. **D**	109. **B**	110. **D**
111. **C**	112. **B**	113. **A,D**	114. **C**	115. **D**	116. **D**	117. **C**	118. **A**	119. **A**	120. **B**
121. **A**	122. **A**	123. **B**	124. **D**	125. **C**	126. **A**	127. **D**	128. **D**		

Chapter 3

REPRODUCTION

1. Plant embryo develops from
   (A) Seed (B) Fruit
   (C) Zygote (D) Flower
2. Embryo of flowering plant is always
   (A) Haploid (n)
   (B) Diploid (2n)
   (C) Triploid (3n)
   (D) Tetraploid (4n)
3. Plant embryo is a mass of
   (A) Cells
   (B) Uncertain tissue
   (C) Collection of plant tissues
   (D) Miniature plant
4. Stamen is a modification of
   (A) Leaf
   (B) Microsporophyll
   (C) Megasporophyll
   (D) Shoot
5. Zygote is formed inside the
   (A) Stigma
   (B) Style
   (C) Female gametophyte
   (D) Seed
6. A microspore is a
   (A) Male gamete
   (B) First cell of male gametophyte
   (C) Last cell of male gametophyte
   (D) Diploid cell
7. An anther consists of
   (A) One microsporangium
   (B) Four microsporangia
   (C) Two microsporangia
   (D) Many microsporangia
8. Cells of nucellus are always
   (A) Haploid (B) Triploid
   (C) Diploid (D) Enucleated
9. The embryo sac is produced from
   (A) Microscope (B) Zygote
   (C) Egg cell (D) Megaspore
10. An egg-apparatus contains
    (A) An egg + two antipodals
    (B) An egg + Secondary nucleus
    (C) An egg + two synergids
    (D) Antipodal cell + synergid
11. In angiosperm, the endosperm nucleus is
    (A) Triploid (B) Diploid
    (C) Tetraploid (D) Haploid
12. Female gametophyte is also known as
    (A) Ovule (B) Egg-apparatus
    (C) Nucellus (D) Embryo sac
13. Embryo sac contains
    (A) 3 eggs (B) 2 eggs
    (C) 1 egg (D) 4 eggs
14. Carpel is formed of
    (A) Two part (B) Three part
    (C) Four part (D) Seven part

15. The arrangement of flowers on the floral axis is known as

(A) Venation (B) Phyllotaxy
(C) Anthology (D) Aestivation

16. The unit of female reproductive body in flower is

(A) Carpel (B) Megasporangium
(C) Ovule (D) Ovary

17. After fertilization, the ovule develops into

(A) Endosperm (B) Seed
(C) Embryo sac (D) Fruit

18. Fruit is a modification of

(A) Female gametophyte
(B) Ovary
(C) Stigma
(D) Nucellus

19. The seed coat develops from

(A) Embryo sac (B) Inner itegument
(C) Nucellus (D) Outer integument

20. Nucellus in seed is represented by

(A) Testa (B) Pericarp
(C) Perisperm (D) Seed coat

21. Pollen grain are produced in

(A) Nucellus (B) Stigma
(C) Anther (D) Chalaza

22. Ovule is attached to the placenta by

(A) Pedicel (B) Hilum
(C) Funicle (D) Petiole

23. Pollen tube enters the embryo sac through

(A) Integument (B) Micropyle
(C) Chalaza (D) Funicle

24. Pollen tube entering in embryo sac has

(A) 3 male gametes
(B) 1 male gametes
(C) 2 male gametes
(D) 4 male gametes

25. In flowering plants, fertilization occur in

(A) Ovary (B) Embryo sac
(C) Nucellus (D) Ovule

26. The formation of zygote without the act of syngamy is called

(A) Polyembryony
(B) Parthenogenesis
(C) Budding
(D) Apospory

27. Which phase of life cycle is dominant in the individuals of angiosperms?

(A) Gametophyte (B) Growth phase
(C) Sporophyte (D) Development phase

28. The process by which seedless fruits are produced are known as

(A) Parthenocarpy
(B) Apogamy
(C) Parthenogenesis
(D) Apospory

29. A flower is

(A) Modified stem
(B) Modified leaf
(C) Modified branch
(D) Modified shoot

30. A flower is specially formed for

(A) Decoration (B) Photosynthesis
(C) Reproduction (D) Fragrance

31. Pollen tube enters the micropyle into

(A) Female gamete
(B) Ovary
(C) Female gametophyte
(D) Nucellus

32. Embryo develops from

(A) Egg cell (B) Zygote
(C) Egg-apparatus
(D) Synergids

33. Fertilized secondary nucleus develops into

(A) Fruit (B) Embryo
(C) Seed (D) Endosperm

34. Transfer of pollen to the stigma is called

(A) Fertilization (B) Germination
(C) Pollination (D) Gametogenesis

35. In ficus, pollination occurs through

(A) Water (B) Air
(C) Bat (D) Insects

36. After fertilization, seed is developed from

(A) Embryo (B) Embryo sac
(C) Ovule (D) Zygote

37. Cross pollination is normally

(A) Not beneficial
(B) Harmful
(C) More beneficial
(D) Rarely seen

38. Pollen grains germinate on

(A) Any surface (B) Stigma
(C) Soil (D) Ovule

39. An anther is also called

(A) Sporangium
(B) Megasporangium
(C) Microsporangium
(D) Stamen

40. The source of food for developing embryo is

(A) Nucellus (B) Ovule
(C) Endosperm (D) Anther

41. Out of megaspore tetrad, the functional megaspore is

(A) Any megaspore
(B) Middle megaspore
(C) Micropylar megaspore
(D) Chalazal megaspore

42. Micropylar end lacks

(A) Egg cell
(B) Synergids
(C) Egg-apparatus
(D) Integument

43. The larger cell of a pollen grain with irregular shaped nucleus is

(A) Apical cell (B) Generative cell
(C) Vegetative cell (D) Basal cell

44. How many megaspore mother cell are produced in a nucleus?

(A) 3 (B) 1
(C) 2 (D) 4

45. How many cells are formed in a mature female gametophyte?

(A) Eight (B) Six
(C) Two (D) Seven

46. Transfer of a pollen grain to the stigma of the same plant is called

(A) Antogamy (B) Geitonogamy
(C) Allogamy (D) Homogamy

47. The uppermost and largest cell of the suspensor which remains in contract with apical cells is called

(A) Hypocotyl (B) Basal cell
(C) Hypophysis (D) Terminal cell

48. The adaptation for self-pollination is

(A) Herkogamy (B) Cleistogamy
(C) Dichrgamy (D) Homogamy

49. Which of the following cell is diploid?

(A) Synergid
(B) Antipodal
(C) Secondary nucleus
(D) Egg cell

50. Suspensor is produced from

(A) Apical cell
(B) Small upper basal
(C) Large Lower basal
(D) Hypophysis cell

51. Which structure pushes the developing embryo toward endosperm to get nutrition?

(A) Hypophysis (B) Terminal octant
(C) Proembryo (D) Suspensor

52. Development of male gametophyte begins
(A) After pollination
(B) Before pollination
(C) On the stigma
(D) In the embryo sac

53. 2 to 3 celled male gametophyte, starts its further development after pollination
(A) In the style (B) In the ovary
(C) On the stigma (D) In the ovule

54. Which part of the male gametophyte disintegrates before fertilization?
(A) Generative nucleus
(B) Tube nucleus
(C) Male gamete (D) Gempore

55. Which of the following is the basal part of ovule?
(A) Placenta (B) Hilum
(C) Micropyle (D) Chalaza

56. In dicot embryogenesis, the first division in zygote is generally
(A) Oblique (B) Longitudinal
(C) Transverse (D) Uncertain

57. Two male gametes are
(A) Produced before pollination
(B) Haploid
(C) Diploid
(D) At the time of pollination

58. The innermost layer of the wall of microsporangium is called
(A) Endothecium (B) Endotermis
(C) Tapetum (D) Intine

59. Pollen grains represent
(A) The future sporophyte
(B) The sporophyte
(C) The gametophyte
(D) The male gametophyte

60. Tapetum provides
(A) Protection to embryo
(B) Nourishment to pollen grains
(C) Nourishment to embryo
(D) Protection to endosperm

61. In triple fusion, how many male gamete participate?
(A) 1 (B) 2
(C) 3 (D) 4

62. Germspores are actually
(A) Apertures in intine
(B) Thick area in intine
(C) Apertures in exine
(D) Thin area in intine

63. Radicle tip is derived from
(A) Suspensor (B) Proembryo
(C) Basal cell (D) Hypophysis

64. How many haploid nuclei are involved in double fertilization?
(A) Four (B) Two
(C) Five (D) Three

65. Endothecium in anther helps in
(A) Dehiscence of anther
(B) Nutrition to pollen
(C) Germination of pollen
(D) Formation of male gamete

66. The intine of a pollen grain is made up of
(A) Lignin and suberin
(B) Pectin and cellulose
(C) Lignin and hemicellulose
(D) Pectin and callose

67. Which is the most resistant natural organic material?
(A) Cellulose (B) Pectin
(C) Suberin (D) Sporpopollenin

68. Style is
(A) A delicate hollow tube
(B) A tough hollow tube
(C) A delicate filament
(D) Called pollen tube

69. Nucellus is mass of
(A) Parenchymatous tissue
(B) Sclerenchymatous tissue
(C) Meristematic tissue
(D) Collenchymatous tissue

70. Which one of the following is the example of mitosis?
(A) Megasporongenesis
(B) Microsporogenesis
(C) Pollen formation
(D) Division of generative cell

71. Typical anther normally consists of
(A) One lobe four sporangia
(B) Two lobe two sporangia
(C) Two lobe four sporangia
(D) One lobe two sporangia

72. During the development of monosporic development of embryo sac, the nucleus of the functional megaspore divides
(A) Two times (B) Three times
(C) One time (D) Repeatedly

73. Cross pollination has an advantage of
(A) Mutation
(B) Polyploidy formation
(C) Genetic recombination
(D) Crossing over

74. Unisexual flowers prevent
(A) Pollination
(B) Breeding
(C) Self-pollination
(D) Cross fertilization

75. The functions of the filiform apparatus is
(A) To nourish the pollen grain
(B) To guide the entry of pollen tube
(C) To develop pollen tube
(D) To carry pollen tube through style

76. Which one of the following is an example of free nuclear endosperm?
(A) Coconout water (B) Castor
(C) Sugarcane juice (D) Groundnut

77. The protective cover of the radicle in maize seed is called
(A) Plumule (B) Coleoptile
(C) Scutelum (D) Coleorhiza

78. In angiosperm, the endosperm is formed
(A) In the nucleus
(B) In the embryo sac
(C) In the seed
(D) In the anther

79. Stamen show homology with
(A) Gametophyte
(B) Male cone
(C) Microsporophyll
(D) Sporangium

80. The pedicel of the female flower coils after pollination in
(A) Lotus (B) Hydrilla
(C) Vallisneria (D) Trapa

81. The arrangement of the haploid nuclei in the normal dicot embryo sac is
(A) 2+3+3 (B) 2+3+2
(C) 3+3+2 (D) 3+2+3

82. In the flowering plants, male and female gametes both are
(A) Motile (B) Non-motile
(C) Diploid (D) Very large

83. Wind pollination requires that the pollen grains are
(A) Heavy and wet
(B) Heavy and non-sticky
(C) Light and dry
(D) Heavy and sticky

84. Future sporophytic generation in a seed is
(A) Cotyledon (B) Endosperm
(C) Hypocotyl (D) Embryo

85. Scutellum

(A) An endosperm
(B) A seed coat
(C) An embryo
(D) A cotyledon

86. Which one of the following floats in the cytoplasm of the vegetative cell?

(A) Male gamete
(B) Generative cell
(C) Female gamete
(D) Microspore

87. The mature pollen grain contains

(A) 3 cells (B) 7 cells
(C) 2 cells (D) 1 cell

88. Pollen tube is formed by

(A) Germ pore (B) Exine
(C) Style (D) Intine

89. A single ovule produces

(A) 3 female gamete
(B) 1 female gamete
(C) 2 female gamete
(D) 4 female gamete

90. Embryo sac is formed inside

(A) Seed (B) Endosperm
(C) Embryo (D) Ovule

91. Proembryo is a

(A) 8 celled structure
(B) 4 celled structure
(C) 2 celled structure
(D) 16 celled structure

92. Suspensor is made up of

(A) 2 to 4 cells
(B) 4 to 8 cells
(C) 8 to 16 cells
(D) 20 to 25 cells

93. Root cap of the embryo develops from

(A) Basal cell (B) Apical cell
(C) Hypophysis (D) Hypocotyl

94. The hilum of the ovule represents the junction between

(A) Nucellus and Embryo
(B) Nucellus and integuments
(C) Funicle and Integuments
(D) Funicle and ovule

95. Which layer of the wall of microsporangium is made up of fibrous layer

(A) Middle layer (B) Endothecium
(C) Tapetum (D) Epidermis

96. Out of the four sets of appendages of a typical flower the outer two sets are

(A) Fertile (B) Reproductive
(C) Sterile (D) Filamentous

97. A proximal sterile part of the stamen is called

(A) Style (B) Connective
(C) Anther (D) Filament

98. A sterile region present between stigma and ovary is called

(A) Pollen tube (B) Style
(C) Filament (D) Suspensor

99. The opposite end of the micropylar region of an ovule is called

(A) Embryo sac (B) Nucellus
(C) Chalaza (D) Thalamus

100. When pollen grains are not ransferred from anthers to stigma in a flower, due to the physical barrier is called

(A) Cleistogamy (B) Herkogamy
(C) Dichogamy (D) Heterogamy

101. The sexual production of seed is called

(A) Fragmentation
(B) Reproduction
(C) Self-fertilization
(D) Dormancy

102. Perisperm is

(A) Peripheral part of endosperm

(B) Remnent of endosperm

(C) Disintegrated secondary nucellus

(D) Persistant of nucellus

103. The root cell of wheat plant has 42 chromosomes. What would be the number of chromosomes in the synergid cell?

(A) 21 (B) 7

(C) 28 (D) 14

104. The plant part which consist of two generations, one within the other, is

(A) Germinated pollen grain

(B) Embryo

(C) Unfertilized ovule

(D) Seed

105. The pollen tube usually enters the female gametophyte

(A) Through one of the synergids

(B) By directly penetrating the egg

(C) Between one synergid and central cell

(D) By knocking off the antipodal cell

106. The male gametic nuclei are the result of mitotic division from which one of the following?

(A) Male gametophyte

(B) Microspore mother cell

(C) Microspores

(D) Generative nucleus

107. The ploidy level in the exine of pollen grain is :

(A) n (B) 3n

(C) 2n (D) 4n

ANSWER SHEET

1. C	2. B	3. D	4. B	5. C	6. B	7. B	8. C	9. D	10. C
11. C	12. B	13. C	14. B	15. C	16. B	17. B	18. B	19. D	20. C
21. C	22. C	23. B	24. C	25. B	26. B	27. C	28. A	29. D	30. C
31. C	32. B	33. D	34. C	35. D	36. C	37. C	38. B	39. C	40. C
41. D	42. D	43. C	44. B	45. D	46. B	47. C	48. B	49. C	50. B
51. D	52. B	53. C	54. B	55. D	56. C	57. B	58. C	59. D	60. B
61. A	62. C	63. D	64. C	65. A	66. B	67. D	68. C	69. A	70. B
71. C	72. B	73. C	74. C	75. B	76. A	77. D	78. B	79. C	80. C
81. B	82. B	83. C	84. D	85. D	86. B	87. C	88. D	89. B	90. D
91. B	92. D	93. C	94. D	95. B	96. C	97. D	98. B	99. C	100. B
101. B	102. D	103. A	104. C	105. A	106. D	107. A			

Chapter 4

FLORAL BIOLOGY

1. The example of protandrous species is
(A) Avocado (B) Walnut
(C) Cabbage (D) Apple

2. A plant bearing both male and female flowers is said to be
(A) Dioecious (B) Monoecious
(C) Polygamous (D) Polycarpic

3. According to position of the ovule within the ovary, an ovule may be
(A) Ascending (B) Descending
(C) Horizontal (D) Vertical

4. An egg or sperm cell in plants is called
(A) Gene (B) Gamete
(C) Zygote (D) Allele

5. An embryo sac of angiosperms generally contains
(A) Two nuclei (B) Three nuclei
(C) Four nuclei (D) Eight nuclei

6. Date palm is
(A) Monoecious (B) Dioecious
(C) Polygamous (D) Gynomonoecious

7. Each stamen is differentiated into
(A) Filament (B) Stigma
(C) Style (D) Ovary

8. Epicalyx is the characteristic of the family
(A) Gramineae (B) Solanaceae
(C) Malvaceae (D) Compositeae

9. Female reproductive organ in the flower is
(A) Calyx (B) Corolla
(C) Stamens (D) Carpels

10. Flower colour in plants is a
(A) Qualitative character
(B) Quantitative character
(C) Complimentary character
(D) Additive character

11. Flower is specialized
(A) Shoot (B) Leaf
(C) Stem (D) Tendril

12. Flowers having both androecium's and gynoecium's are said to be
(A) Hermaphrodite (B) Unisexual
(C) Staminate (D) Pistilate

13. Fusion of the one sperm (male gamete) and definitive nucleus results in the formation of
(A) Endosperm (B) Embryo
(C) Saprophyte (D) Zygote

14. Generally embryosac is
(A) Monosporic (B) Bisporic
(C) Tetrasporic (D) Octosporic

15. Gynoecium or pistil in the female whorl differentiated into
(A) Stamen (B) Style
(C) Anther (D) Androecium

16. In angiosperm
(A) Sporophyte is complex and gametophyte inconspicuous
(B) Sporophyte simple and gametophyte complex
(C) Sporophyte and gametophyte complex
(D) Sporophyte and gametophyte inconspicuous

17. In China, rose flower is
(A) Hypogynous (B) Perigynous
(C) Epigynous (D) Semigynous

18. In compositeae, the gynoecium is syncarpous bicarpellary with
(A) Marginal placentation
(B) Axile placentation
(C) Free central placentation
(D) Basal placentation

19. In cruciferae, the stamens are said to be
(A) Didynamous
(B) Tridynamous
(C) Tetradidynamous
(D) Decadynamous

20. In family malvaceae, the stamens are
(A) Diadelphous (B) Gynandrous
(C) Syngenesious (D) Monoadelphous

21. In Ficus (banyan, pepal, fig), the inflorescence is
(A) Cyathuim (B) Verticillaster
(C) Hypanthodium (D) Spadix

22. In potato, the inflorescence is
(A) Helicoid (B) Scropioid cyme
(C) Umbel (D) Raceme

23. In *Primula* sp
(A) Sporophytic incompatibility found
(B) Gametophytic incompatibility found
(C) Heteromorphic incompatibility found
(D) Homomorphic incompatibility found

24. In self-incompatibility system
(A) Pollen not produced
(B) Self pollination takes place
(C) Pollen does not germinate on self-pollination
(D) Cross pollination not take place

25. In solanaceae, the flowers are
(A) Actionomorphic bisexual and epigynous
(B) Zygomorphic, unisexual and perigynous
(C) Actinomorphic, bisexual and hypogynous
(D) Actinomorphic, bisexual and perigynous

26. Which one of the following is not dioecious plant?
(A) Datepalm (B) Vallisnaria
(C) Maize (D) Papaya

27. In the cymose inflorescence, the flowers develop in
(A) Basipetal succession
(B) Acropetal succession
(C) Centripetal succession
(D) Axilepetal succession

28. In the ovary of marigold the placentation is
(A) Superficial (B) Basal
(C) Marginal (D) Axile

29. In the simple ovary (of one carpel), there is one common type of placentation known as
(A) Axile (B) Parietal
(C) Central (D) Marginal

30. Maize is
(A) Monoecious plant
(B) Dioecious plant
(C) Polygamous
(D) None of the above

31. Megaspore mother cell produces four megaspores out of which usually

(A) Only one megaspore is functional and the and remaining three megaspores abort

(B) Two megaspores are functional the remaining two megaspores abort

(C) Three megaspores are functional and the fourth megaspore aborts

(D) All the four megaspores are functional

32. ----------------estimated that incompatibility occurs in more than 3,000 species among 20 families of the flowering plants

(A) D.De Winton

(B) M.B. Hughes

(C) W.H. Gabelman

(D) E.M. East

33. Pea has stamens

(A) Diadelphous (B) Monoadelphous

(C) Polyadelphous (D) Gynandrous

34. Pointed gourd is a

(A) Monoecious plant

(B) Hermaphrodite plant

(C) Dioecious plant

(D) Polygamous plant

35. Spikelets are characteristic of the family

(A) Umbelliferae (B) Gramineae

(C) Malvaceae (D) Leguminoceae

36. The commonest form of the ovule is

(A) Orthotropous

(B) Anatropous

(C) Amphilotropous

(D) Campylotropous

37. The group of plants, which bear flowers and seeds, are known as

(A) Crytogams (B) Phanerogams

(C) Pteridophyta (D) Prokaryate

38. The haploid phase in plants is called

(A) Sporophyte (B) Gametophyte

(C) Zygote (D) Embryo sac

39. The inflorescence of banana is known as

(A) Catkin (B) Raceme

(C) Spike (D) Spadix

40. The main body of ovule is called

(A) Hilum (B) Chalaza

(C) Nucellus (D) Integuments

41. When the pistil has all the carpels united together, it is said to be

(A) Apocarpous (B) Syncarpous

(C) Simpler types (D) Complex types

42. The mode of arrangement of ovules in the cavity of ovary is known as

(A) Pollination (B) Aestivation

(C) Placenatation (D) Fertilization

43. Which is the monoecious plant?

(A) Maize (B) Datepalm

(C) Rice (D) Mango

44. The number of stamens is indefinite in

(A) Cruciferae (B) Solanaceae

(C) Papillionaceae (D) Malvaceae

45. The ovary of family liliaceae is

(A) Trilocular superior with axile placentation

(B) Trilocular inferior with parietal placenation

(C) Trilocular inferior with basal placentation

(D) Trilocular superior with free central placentation

46. Which is the dioecious plant?

(A) Maize (B) Marigold

(C) Vallisneria (D) Rice

47. Umbel inflorescence is characteristic of the family

(A) Mustard family (B) Okra family

(C) Pea family (D) Coriander family

48. Unisexual flowers bear
(A) Only stamens
(B) Only carpels
(C) Either stamens or carpels
(D) Both stamens and carpels

49. When the stamens are united throughout their whole length by both the filaments and the anthers, they are said to be
(A) Syngenesious stamens
(B) Polyadelphous stamens
(C) Synandrous stamens
(D) Diadelphous stamens

50. Veticillaster inflorescence is characteristic of family
(A) Malvaceae (B) Pedaliaceae
(C) Liliaceae (D) Habitatace

51. When stamens are united with the carpels, stamens are said to be
(A) Synandrous (B) Syngenesious
(C) Gynandrous (D) Syngamous

52. When the anthers are united into a bundle or tube, but the filaments are free, the stamens are said to be
(A) Syngenesious
(B) Synandrous
(C) Monoadelphous
(D) Polyadelphous

53. When the ovule bends along the funicle that the micropyle lays close to the hilum, the ovule is said to be
(A) Orthotropous
(B) Anatropous
(C) Amphilotropous
(D) Campylotropous

54. When the ovule is bent round like a horseshoe, it is said to be
(A) Orthotropous
(B) Anatropous
(C) Amphilotropous
(D) Campylotropous

55. Plants with bisexual flowers include
(A) Custard apple
(B) Sage
(C) Tulip
(D) Both A and B

56. These are the ovule of a plant that has been fertilized by pollen that can grow into a new plant.
(A) Seeds (B) Roots
(C) Stems (D) Leaves

57. Embryo sac is present inside the
(A) Ovule (B) Stigma
(C) Micropyle (D) Style

58. The number nuclei present inside the embryo sac are
(A) 4 (B) 2
(C) 5 (D) 3

59. It is an indeterminate inflorescence with pedicellate flowers along the axis. What is that inflorescence?
(A) Spike (B) Spadix
(C) Raceme (D) Capitulum

60. It is a type of inflorescence that do not have a pedicel. What it is?
(A) Spike (B) Spadix
(C) Raceme (D) Capitulum

61. What is the type of flower in sunflower?
(A) Spike (B) Spadix
(C) Raceme (D) Capitulum

62. What is the type of flower in anthurium?
(A) Spike (B) Spadix
(C) Raceme (D) Capitulum

63. Which one of the following is the female part of a flower?
(A) Calyx (B) Corolla
(C) Stamen (D) Pistil

64. Which one of the following is the male part of a flower?
(A) Calyx (B) Corolla
(C) Stamen (D) Pistil

65. Which is the enlarged portion of the pistil?
(A) Anther (B) Ovary
(C) Filament (D) Ovule

66. What do you name, if the ovules are positioned on the central axis of ovary?
(A) Axile placentation
(B) Marginal placentation
(C) Basal placentation
(D) Parietal placentation

67. What do you name, if the ovules are arranged at the base of the ovary?
(A) Axile placentation
(B) Marginal placentation
(C) Basal placentation
(D) Parietal placentation

68. What do you call, if the ovules are arranged at the roof of the ovary?
(A) Apical placentation
(B) Marginal placentation
(C) Basal placentation
(D) Parietal placentation

69. -----------------, the stamens are sometimes attached to the petals or the gynoecium
(A) Antipetalous
(B) Antisepalous
(C) Diplostemonous
(D) Epipetalous

70. What is the type of gynoecium which has a single carpel?
(A) Apocarpous (B) Monocarpous
(C) Syncarpous (D) Epipetalous

71. A flower is having both androecium and gynoecium. What is the type of flower?
(A) Hermaphrodite flower
(B) Imperfect flower
(C) Staminate flower
(D) Pistillate flower

72. A flower is having an androecium only, but no gynoecium. What is the type of flower?
(A) Hermaphrodite flower
(B) Imperfect flower
(C) Staminate flower
(D) Pistillate flower

73. A flower is possessing a gynoecium only, but no androecium. What is the type of flower?
(A) Hermaphrodite flower
(B) Imperfect flower
(C) Staminate flower
(D) Pistillate flower

74. A flower is having a reproductive structure that is either functionally male or functionally female. Which one of the following is correct for this statement?
(A) Hermaphrodite flower
(B) Imperfect flower
(C) Staminate flower
(D) Pistillate flower

75. ------------------ is an individual that has both androecium and gynoecium reproductive units.
(A) Monoecious (B) Dioecious
(C) Androecious (D) Gynoecious

76. -------------- refers to a plant population having separate staminate and pistillate plants.
(A) Monoecious (B) Dioecious
(C) Androecious (D) Gynoecious

77. ---------------- is a plant that produces male flowers only, but does not produce seed.
(A) Monoecious
(B) Dioecious
(C) Androecious
(D) Gynoecious

78. ----------------- is a plant that possesses female flowers only and produces seed, but no pollen.
(A) Monoecious (B) Dioecious
(C) Androecious (D) Gynoecious

79. The filaments are fused together to form one group and anthers are free. What it is called?
(A) Monadelphous (B) Diadelphous
(C) Polyadelphous (D) Syngenesious

80. Filaments are fused to form two bundles as in pea. What is it?
(A) Monadelphous (B) Diadelphous
(C) Polyadelphous (D) Syngenesious

81. Filaments are fused to form many bundles as in lemon and it is known as
(A) Monadelphous
(B) Diadelphous
(C) Polyadelphous
(D) Syngenesious

82. Filaments are free and anthers are free as in sunflower and it is known as
(A) Monadelphous
(B) Diadelphous
(C) Polyadelphous
(D) Syngenesious

83. Which of the following are two nuclei inside the embryo sac
(A) Egg cell and secondary nucleus
(B) Micropile and Egg cell
(C) Stamen and Stigma
(D) Anther and Style

84. The tip of the pollen tube breaks when the pollen tube passes into the ovule through
(A) Micropyle (B) Embryo
(C) Calyx (D) Corolla

85. Fragnant flowers with well developed nectaries are adapted for
(A) Anemophily (B) Hydrophily
(C) Entomophily (D) Zoophily

86. Generative nucleus divides forming
(A) Three female nuclei
(B) Two male nuclei
(C) Three male nuclei
(D) Two female nuclei

87. The number of outer layers present in each ovule are
(A) 2 (B) 3
(C) 4 (D) 5

88. At the upper end of the outer layer of embryosac there is an opening known as
(A) Stigma (B) Style
(C) Micropyle (D) Anther

89. A monocarpic plant is one which
(A) Has only one carpel in lifetime and thereafter dies
(B) Flowers and fruits only once
(C) Produces only one seed
(D) Has two carpels

90. Vernalisation stimulates flowering in
(A) Carrot (B) Turmeric
(C) Elephant Yam (D) Ginger

91. Hypanthodium refers to a special type of
(A) Fruit (B) Flower
(C) Inflorescence (D) Ovary

92. The number of open present in each ovule are
(A) 1 (B) 3
(C) 4 (D) 5

93. At the upper end of the outer integument there is an opening known as
(A) Stigma (B) Style
(C) Micropyle (D) Anther

94. Ovule sac is present inside the
(A) Ovary (B) Stigma
(C) Micropyle (D) Style

95. The number male gametes present inside the pollen tube are

(A) 4 (B) 2
(C) 5 (D) 3

96. The arrangement of the nuclei in a normal embryo sac in the plants is

(A) 2+3+3 (B) 3+2+3
(C) 2+4+2 (D) 3+3+2

97. Fibrous thickenings of hygroscopic nature are found in this part of the anther wall

(A) Middle layers (B) Endothecium
(C) Epidermis (D) Tapetum

98. Ovule is attached to the placenta of ovary wall by

(A) Funicle (B) Raphae
(C) Hilum (D) Testa

99. Sporopollenin is chemically

(A) Protein
(B) Fatty substance
(C) Homopolysaccharide
(D) Heteropolysaccharide

100. The gynoecium consists of many free pistils in flowers of

(A) Papaver (B) Michelia
(C) Aloe (D) Tomato

101. In anatropous ovule, what is the direction of micropyle?

(A) Right (B) Downward
(C) Upward (D) Left

102. When the other floral parts are arranged at the base of the gynoecium, the flower is called

(A) Perigynous flower
(B) Epigynous flower
(C) Hypogynous flower
(D) Polygynous flower

103. An orthotropous ovule is one in which micropyle and chalaza are

(A) In straight line of funicle
(B) Parallel along with funicle
(C) At right angle to funicle
(D) Parallel along with ovule

104. Cauliflory is flowers produced

(A) An old dormant bud
(B) On young bud
(C) On the axil
(D) On the branch

105. Date palm is :

(A) Monoecious (B) Polygamous
(C) Dioecious (D) Bisexual

106. Cliestogamy encourages :

(A) Self pollination
(B) Apomixis
(C) Cross pollination
(D) Inbreeding depression

107. Epicalyx is the characteristic of the family :

(A) Gramineae (B) Malvaceae
(C) Solanaceae (D) Compositeae

108. Embryo sac is :

(A) Male gametophyte
(B) Sporophyte
(C) Microgametophyte
(D) Female gametophyte

109. Sex from in most cucurbits including melons is :

(A) Monoecious (B) Dioecious
(C) Hermaphrodite (D) Bisexual

110. Each mature anther has :

(A) One pollen sac
(B) Two pollen sacs
(C) Three pollen sacs
(D) Four pollen sacs

111. In *Primula* sp :

(A) Sporophytic incompatibility found
(B) Gametophytic incompatibility found
(C) Homomorphic incompatibility found
(D) Heteromorphic incompatibility found

112. In potato, the inflorescence is :

(A) Raceme (B) Scropioid cyme
(C) Umbel (D) Helicoid

113. In Ficus (banyan, pepal, fig), the inflorescence is :

(A) Cyathuim (B) Verticillaster
(C) Spadix (D) Hypanthodium

114. In family malvaceae, the stamens are :

(A) Monoadelphous
(B) Diadelphous
(C) Syngenesious
(D) Gynandrous

115. In compositeae, the gynoecium is syncarpous bicarpellary with :

(A) Basal placentation
(B) Free central placentation
(C) Axile placentation
(D) Marginal placentation

116. Orange or lemon fruit is known as :

(A) Pome (B) Pepo
(C) Balusta (D) Hesperidium

117. Pea has stamens :

(A) Monoadelphous
(B) Diadelphous
(C) Polyadelphous
(D) Gynandrous

118. Secondary nucleus in the embryo sac is :

(A) Haploid (B) Diploid
(C) Triploid (D) Hexaploid

119. Spikelets are characteristic of the family :

(A) Leguminoceae (B) Gramineae
(C) Umbelliferae (D) Malvaceae

120. Sporogenesis means :

(A) Formation of zygote
(B) Formation of embryo
(C) Germination of seed
(D) Development and formation of spores

121. According to position of the ovule within the ovary, an ovule may be :

(A) Suspended (B) Pendulous
(C) Ascending (D) All of the above

122. Gynoecium or pistil in the female whorl differentiated into :

(A) Filament (B) Stamen
(C) Anther (D) Pollen tube

123. Female reproductive organ in the flower is:

(A) Calyx (B) Carpels
(C) Stamens (D) Corolla

124. A male gametophyte in angiosperms contain :

(A) One gametes only
(B) Two male gametes only
(C) Two male gametes and one tube nucleus
(D) One male gamete and one tube nucleus

125. In sunflower, incompatibility is :

(A) Sporophytic
(B) Gametophytic
(C) Sporophytic and gametophytic
(D) Sporophytic or gametophytic

126. The tissue which nourish microspores is :

(A) Nucellus (B) Mesophyll
(C) Tapatum (D) Endodermis

127. The stalk of the ovules is called :

(A) Pedicel (B) Micropyle
(C) Funicle (D) Petiole

128. Flowers having both androecium and gynoecium are said to be :

(A) Staminate (B) Pistilate
(C) Hermaphrodite (D) Unisexual

129. In cruciferae, the stamens are said to be:

(A) Didynamous
(B) Tridynamous
(C) Tetradidynamous
(D) Decadynamous

130. Each megaspore mother cell finally produces :

(A) Only one megaspore
(B) Two megaspores
(C) Four megaspores
(D) Several megaspores

131. Flower is specialized :

(A) Shoots (B) Leaf
(C) Tendril (D) Stem

132. Each stamen is differentiated into :

(A) Style (B) Stigma
(C) Ovule (D) Anther

ANSWER SHEET

1. **B**	2. **B**	3. **A**	4. **B**	5. **D**	6. **B**	7. **A**	8. **C**	9. **D**	10. **A**
11. **A**	12. **A**	13. **A**	14. **B**	15. **B**	16. **A**	17. **A**	18. **C**	19. **C**	20. **D**
21. **C**	22. **A**	23. **C**	24. **C**	25. **C**	26. **C**	27. **A**	28. **B**	29. **D**	30. **A**
31. **A**	32. **D**	33. **A**	34. **C**	35. **B**	36. **B**	37. **B**	38. **B**	39. **D**	40. **C**
41. **B**	42. **C**	43. **A**	44. **D**	45. **A**	46. **C**	47. **D**	48. **C**	49. **C**	50. **D**
51. **C**	52. **A**	53. **B**	54. **C**	55. **A**	56. **A**	57. **A**	58. **B**	59. **C**	60. **A**
61. **D**	62. **B**	63. **D**	64. **C**	65. **B**	66. **A**	67. **C**	68. **A**	69. **D**	70. **B**
71. **A**	72. **C**	73. **D**	74. **B**	75. **A**	76. **B**	77. **C**	78. **D**	79. **A**	80. **B**
81. **C**	82. **D**	83. **A**	84. **A**	85. **C**	86. **B**	87. **A**	88. **C**	89. **B**	90. **A**
91. **C**	92. **A**	93. **C**	94. **A**	95. **B**	96. **B**	97. **B**	98. **A**	99. **B**	100. **B**
101. **B**	102. **C**	103. **A**	104. **D**	105. **C**	106. **A**	107. **B**	108. **D**	109. **A**	110. **D**
111. **D**	112. **D**	113. **D**	114. **A**	115. **B**	116. **D**	117. **B**	118. **B**	119. **B**	120. **D**
121. **D**	122. **A**	123. **B**	124. **C**	125. **A**	126. **C**	127. **C**	128. **C**	129. **C**	130. **A**
131. **B**	132. **D**								

Chapter 5

POLLINATION

1. How synchronous flowering is useful in hybrid seed production?

   (A) Enhances pollen fertility and poor seed set
   (B) Enhances pollen viability and seed viability
   (C) Enhances pollen dispersal and pollen storage
   (D) Enhances fertilization and seed set

2. Non-synchronous flowering is observed in the hybrid seed production of

   (A) Bajra (B) Sorghum
   (C) Redgram (D) Sunflower

3. The classical example for protandry is

   (A) Maize (B) Wheat
   (C) Groundut (D) Cotton

4. ------------ is a classical example for protogyny

   (A) Sorghum (B) Pearl millet
   (C) Greengram (D) Sesamum

5. Which one of the following crop is a self pollinated crop?

   (A) Tomato (B) Bitter gourd
   (C) Cotton (D) Turmeric

6. Which one of the following crop is a cross pollinated crop?

   (A) Tomato (B) Bitter gourd
   (C) Cotton (D) Turmeric

7. Which one of the following crop is a often cross pollinated crop?

   (A) Tomato (B) Bitter gourd
   (C) Cotton (D) Turmeric

8. Which one of the following crop does not involve pollination at all?

   (A) Tomato (B) Bittergourd
   (C) Cotton (D) Turmeric

9. ------------- is an example for protogyny

   (A) Maize (B) Bajra
   (C) Rice (D) Wheat

10. Bajra (Pearl millet) belongs to which type of anthesis?

    (A) Autogamy (B) Protogyny
    (C) Allogamy (D) Dichogamy

11. In which one of the family, the protogyny mechanism exist?

    (A) Malvaceae (B) Convolvulaceae
    (C) Magnoliaceae (D) Leguminosae

12. In which one of the family, the protandry occurs?

    (A) Malvaceae (B) Convolvulaceae
    (C) Magnoliaceae (D) Laminaceae

13. ------------ is a mechanism in which both sex organs mature at the same time

    (A) Homogamy (B) Cleistogamy
    (C) Allogamy (D) Autogamy

14. ---------------- is an example for homogamy

(A) Maize (B) Pearl millet
(C) Mirabilis (D) Sapota

15. ------------ is an example for cleistogamy

(A) Maize (B) Pearl millet
(C) Mirabilis (D) Arachis

16. ------------- is an example for chasmogamy

(A) Maize (B) Rice
(C) Mirabilis (D) Arachis

17. ---------------- avoids cross pollination

(A) Hidden style and ovary
(B) Hidden filament and pedicel
(C) Hidden stamen and stigma
(D) Hidden corolla and calyx

18. ---------------- closely surrounded by anthers which helps in self pollination

(A) Calyx (B) Corolla
(C) Filament (D) Stigma

19. ------------- is an example for hydrophily

(A) Maize (B) Mango
(C) Hydrilla (D) Water hyacinth

20. ------------- is an example for anemophily

(A) Rice (B) Wheat
(C) Coconut (D) Mango

21. Which one of the following crop is an example for entomophily?

(A) Groundnut (B) Sesamum
(C) Wheat (D) Sunflower

22. Which one of the following crop is an example for ornithophily?

(A) Erythrina (B) Bombax
(C) Mango (D) Coconut

23. ------------- is an example for pollination by beetles

(A) Erythrina (B) Bombax
(C) Mango (D) Coconut

24. ------------- is an example for pollination by butterflies

(A) Erythrin (B) Bombax
(C) Mango (D) Castor

25. ------------- is an example for pollination by bats

(A) Mango (B) Kapok
(C) Arecanut (D) Coconut

26. In which crop, heterostyly mechanism exits?

(A) Brinjal (B) Tomato
(C) Chillies (D) Bhendi

27. In which crop, self-incompatibility occurs

(A) Brinjal (B) Tomato
(C) Sunflower (D) Sesamum

28. An example for andromonoecious crop is

(A) Brinjal (B) Tomato
(C) Sunflower (D) Sesamum

29. An example for gynomonoecious crop is

(A) Brinjal (B) Tomato
(C) Sunflower (D) Ailanthus

30. An example for trimonoecious crop is

(A) Brinjal (B) Banana
(C) Sunflower (D) Ailanthus

31. Which of the following contains reserve food material

(A) Stamen (B) Stigma
(C) Endosperm (D) Anther

32. After fertilization, ovules develop into

(A) Thalamus (B) Seeds
(C) Predicel (D) Fruits

33. After fertilization, ovary develops into

(A) Seeds (B) Thalamus
(C) Fruit (D) Pedicel

34. Transfer of pollen grains from the anther to the stigma of the flower of the same species is called

(A) Pollination (B) Fertilization
(C) Adoption (D) Diffusion

35. Small and stout stalk of Datura flower at its base is called

(A) Sepals (B) Pedicel
(C) Thalamus (D) Corolla

36. Pollination by wind takes place in

(A) Maize (B) Paddy
(C) Grass (D) All the above

37. In Vallisneria, Hydrilla Pollination takes place by

(A) Wind (B) Water
(C) Insects (D) All the above

38. Flowers like night queen, jasmine emit a sweet scent to attract insects like

(A) Honey bees
(B) Butterflies
(C) Beetles and moths
(D) All the above

39. The fusion of the male reproductive nucleus and the female reproductive nucleus is called

(A) Fertilization (B) Adoption
(C) Regeneration (D) Excretion

40. At the end of pollen tube there are two nuclei called

(A) Sperm and ovum
(B) Tube nucleus and generative nucleus
(C) Tube nucleus and sperm
(D) Generative nucleus and stigma

41. The pollen grains of a flower are transferred to the stigma of the same flower in

(A) Diffusion (B) Cross pollination
(C) Self pollination (D) Fertilization

42. Self pollination occurs in plants like

(A) Wheat (B) Tobacco
(C) Peas (D) All the above

43. Pollen from one flower is carried to the stigma of another flower of the same plant or a different plant of same species in

(A) Self pollination
(B) Cross pollination
(C) Fertilization
(D) Diffusion

44. Cross pollination takes place in

(A) Tomato
(B) Hibiscus and ladies finger
(C) Brinjal
(D) All the above

45. Which of the following help in cross pollination in unisexual flowers?

(A) Wind (B) Bees
(C) Butterflies (D) All the above

46. The swollen part over the pedicel is

(A) Thalamus
(B) Corolla
(C) Androecium
(D) Gynoecium

47. The parts of the thalamus are

(A) Calyx and Corolla
(B) Androecium
(C) Gynoecium
(D) All the above

48. The outermost whorl consisting of five green leaf structures of Datura flower is called

(A) Petals (B) Calyx
(C) Sepals (D) Stamens

49. All the sepals together are known as

(A) Stamens (B) Calyx
(C) Corolla (D) Petals

50. The whorl lying inner to the calyx is called

(A) Stamens (B) Stigma
(C) Androecium (D) Corolla

51. The Corolla which consists a number of delicate structures called

(A) Calyx (B) Stamens
(C) Petals (D) Sepals

52. What is the dominating colour of petals in cotton
(A) Blue (B) Red
(C) Yellow (D) Green

53. The long thin structures present inside the corolla are called
(A) Sepals (B) Stamens
(C) Style (D) Filament

54. The number of stamens present in Datura are
(A) 5 (B) 9
(C) 7 (D) 8

55. Stamens are the male reproductive parts of the flower and are collectively known as the
(A) Anther (B) Androecium
(C) Gynoecium (D) Filament

56. Stalk of the stamen is called the
(A) Anther (B) Filament
(C) Style (D) Gynoecium

57. The expanded part a tip of the stamen is called
(A) Androecium (B) Pistil
(C) Anther (D) Pollen grains

58. Which of the following produce, fine particles called pollen or pollen grains?
(A) Filament (B) Style
(C) Anther (D) Stigma

59. Which of the following contain male gametes?
(A) Overy (B) Pistil
(C) Style (D) Pollen grains

60. Gynoecium is also known as
(A) Pistil (B) Androecium
(C) Stigma (D) Style

61. The base of the pistil is known as
(A) Ovary (B) Stigma
(C) Style (D) Calyx

62. A thin long structure starting from ovary is
(A) Pistil (B) Stigma
(C) Style (D) Ovary

63. The tip of the style is known as
(A) Corolla (B) Stigma
(C) Calyx (D) Pistil

64. Which of the following has three parts namely ovary, style and stigma
(A) Pistil (B) Androecium
(C) Gynoecium (D) Calyx

65. The grain like structures in the ovary are called as
(A) Calyx (B) Pistil
(C) Ovules (D) Style

66. Which of the following contain female gametes?
(A) Stigma (B) Pistil
(C) Calyx (D) Ovules

67. Which of the following play an important role in the sexual reproduction of flower?
(A) Calyx
(B) Pistil
(C) Androecium and Gynoecium
(D) Stigma

68. Which of the following protect the Androecium and gynoecium of the flower?
(A) Pedicel (B) Calyx and corolla
(C) Stigma (D) Style

69. Which of the following are known as accessory parts of the flower?
(A) Stigma (B) Androecium
(C) Calyx and Corolla
(D) Gynoecium

70. Which of the following are productive parts of the flower?
(A) Calyx (B) Corolla
(C) Pedicel
(D) Andoecium and gynoecium

71. In the angiosperm life cycle, which of the following is not a member of the

(A) Carpel (B) Anther

(C) Leaf (D) Ovary

72. A number of megaspores degenerate after being produced by the meiosis of the megaspore mother cell. That number is

(A) 4 (B) 3

(C) 2 (D) 1

73. Early seed plants are pollinated by

(A) Bees (B) Birds

(C) Wind (D) Butterflies

74. A plant that flowers in its second year of growth and then dies is

(A) A biennial (B) A semi-annual

(C) An annual (D) A perennial

75. Which of the following flowering arrangements promotes self-pollination?

(A) Monoecious (B) Dioecious

(C) Dichogamous (D) None of the above

76. Herbaceous perennials store food in their

(A) Branches (B) Trunk

(C) Roots (D) Shoots

77. Which of the following is not an annual plant?

(A) Cabbage (B) Corn

(C) Sunflower (D) Wheat

78. The male gametophyte of a flowering plant is the

(A) Pollen sac (B) Microspore

(C) Sperm cell (D) Megaspore

79. If a flower is to be pollinated by moths, it should be

(A) White (B) Blue

(C) Purple (D) Orange

80. Plants that produce only ovules or pollen are called

(A) Monoecious (B) Monogamous

(C) Dichogamous (D) Dioecius

81. Angiosperm fossil pollen appears clearly in the early part of the ------------ period.

(A) Triassic (B) Permian

(C) Carboniferous (D) Jurassic

82. Which of the following is a plant whose seeds are dispersed by wind?

(A) Elm (B) Oak

(C) Willow (D) Cottonwood

83. If a flower is to be pollinated by bees, it should be

(A) Red (B) White

(C) Blue (D) Green

84. In most plants, petals are most similar to

(A) Leaves (B) Stamens

(C) Sepals (D) Carpels

85. Angiosperms are believed to have evolved in a climate that was characterized by ------ conditions.

(A) Extreme temperatures and dry

(B) Even temperatures and moist

(C) Warm temperatures and dry

(D) Cold and dry

86. Plants producing both staminate and pistilate flower on the same plant are said to be

(A) Monoecius (B) Dioecious

(C) Heteroecious (D) Dichagmous

87. A trend in floral evolution is

(A) Bilateral symmetry

(B) Spiral symmetry

(C) Radial symmetry

(D) Lateral symmetry

88. Genetic self-incompatibility tends to increase

(A) Pollen dispersal

(B) Inbreeding

(C) Self pollination

(D) Cross pollination

89. Many genes are known that affect both leaves and

(A) Petals (B) Stamens
(C) Sepals (D) Carpels

90. If a flower is to be pollinated by birds, it should be

(A) Blue (B) White
(C) Yellow (D) Orange

91. The female gametophyte of an angiosperm is the

(A) Embryo sac
(B) Ovule
(C) Megaspore mother cell
(D) Egg cell

92. If a flower is to be pollinated by moths, it should

(A) Be heavily scented
(B) Be shaped like a tulip
(C) Close about noon
(D) Be brightly colored

93. Which of the following is not characteristic of primitive flowers?

(A) Flowering parts are numerous
(B) Flowers are arranged in definite whorls
(C) Petals and sepals are similar
(D) Unfused flower parts

94. When one plants potatoes, the seed pieces are parts of

(A) Stolons (B) Adventitious
(C) Rhizomes (D) Suckers

95. Which of the following is not a major adaptation that has contributed to the enormous success of angiosperms over all other vascular plants?

(A) Seeds (B) Flowers
(C) Fruit (D) Racheids

96. The part of a flower that protects the developing bud is the

(A) Ovary (B) Petal
(C) Sepal (D) Anther

97. The part of the flower that produces pollen is the

(A) Ovary (B) Petal
(C) Sepal (D) Anther

98. Which of the following is not part of the innermost whorl of the flower?

(A) Stamens (B) Style
(C) Pistil (D) Stigma

99. Pollen grains on the stigma of a flower reach the egg by forming a

(A) Cotyledon (B) Seed
(C) Pollen tube (D) Carpel

100. Self-pollination can occur in which of the following

(A) Bisexual plants
(B) Dioecious plants
(C) Pistillate plants
(D) Staminate plants

101. Angiosperms produce ovules that are completely enclosed by sporophyte tissue at the time they are pollinated

(A) True (B) False

102. Pollination by insects depends on the random movement of insects from flower to flower

(A) True (B) False

103. In angiosperms, the male gametophyte contains a single sperm that unites with the egg to produce the seed.

(A) True (B) False

104. When pollen from the same flower is brought to the stigma of a flower from insects, wind, etc., it is called

(A) Pollination
(B) Fertilization
(C) Germination
(D) Anthesis

105. When the pollen unites with the ovule to form a seed it is called

(A) Pollination (B) Fertilziation
(C) Germination (D) Anthesis

106. Flowers that have bright colored petals rather than dull petals are pollinated by

(A) Insects (B) Wind
(C) Humans (D) Rain

107. Examples of flowers that are pollinated by wind are

(A) Buttercup and rose
(B) Sunflower
(C) Orchid and wall flower
(D) Corn and willow

108. Flowers that have small pollen grains are pollinated by

(A) Humans (B) Anthers
(C) Insects (D) Wind

109. Types of pollination includes

(A) Cross pollination
(B) Self pollination
(C) Self fertilization
(D) Both A and B

110. Considering wind pollinated and insect pollinated flowers, flowers that are pollinated by wind

(A) Have pinhead shape branches
(B) Have upward facing flowers
(C) Do not produce nectar
(D) Are large in size

111. Pollination is best defined as

(A) Germination of pollen grains
(B) Visiting flowers by insects
(C) Transfer of pollen from anther to stigma
(D) Growth of pollen tube in ovule

112. Pollinia are sac like structures

(A) In which anther lobes are present
(B) Which are present in megasporangia
(C) In which pollen grains are present in mass
(D) Which secrete yellow substance called pollenkit material

113. Intine of pollen grain composed of

(A) Lipid and protein
(B) Cellulose and pectin
(C) Pectin and lignin
(D) Lignin and cutin

114. The exine of a pollen grain is made of

(A) Pectin and cellulose
(B) Sporopollenin
(C) Pollenkit
(D) Lignocellulose

115. Pollination which occurs in closed flower is known as

(A) Allogamy
(B) Cleistogamy
(C) Protogyny
(D) Syngamy

116. Insect pollinated flowers usually possess

(A) Dry pollens with smooth surface
(B) Sticky pollens with rough surface
(C) Large quantities of pollen
(D) Brightly coloured pollens

117. Pollen grains of flowers pollinated by insects or wind are not

(A) Large and showy
(B) Rough and sticky
(C) Smooth and dry
(D) Rough and Dry

118. The deposition of pollen on the stigma of another flower of the same plant is known as

(A) Homogamy
(B) Dichogamy
(C) Geitonogamy
(D) Xenogamy

119. Self pollination occurs between stamens and carpels of the
(A) Same flower
(B) Different flowers
(C) Same flower or the different flowers of the same plant
(D) Different flowers of the same plant

120. If a pollen of a flower falls on the stigma of another flower belonging to the same plant it is
(A) Genetically self pollination and ecologically cross pollination
(B) Ecologically cross pollination
(C) Genetically and ecologically cross pollination
(D) None of these

121. A typical example of self pollination is
(A) Wheat (B) Rice
(C) Cotton (D) Maize

122. Pollenkit is present in the pollen wall of
(A) Anemophilous flowers
(B) Entomophilous flowers
(C) Zoophilous flowers
(D) Malacophilous flowers

123. Wind pollinated plants differ from insect pollinated plants in having
(A) Small petals and sticky pollen
(B) Small coloured petals and heavy pollen
(C) Coloured petal and large pollen
(D) No petals and light pollen

124. Pollination by bats is called
(A) Ornithophily (B) Entomophily
(C) Malacophily (D) Cheiropterophily

125. Pollination is characteristic of
(A) Gymnosperms (B) Fungi
(C) Angiosperms (D) Pteridophytes

126. Offsprings receive valuable qualities from both parents in
(A) Cross pollination
(B) Self pollination
(C) Wind pollination
(D) Insect pollination

127. In order to restore vigor of species, pollination must be
(A) Self pollination
(B) Cross pollination
(C) Wind pollination
(D) Insect pollination

128. In custard apple,
(A) Self pollination is highly likely
(B) Flowers are dioecious
(C) Anthers and stamens may mature at different times
(D) Flowers are monecious

129. In plants, where self-pollination is unlikely due to greater distance between male and female parts includes
(A) Pawpaw (B) Custard apple
(C) Sage (D) Rose

130. Distance between stigma and anthers in same flower,
(A) Reduces chances of self pollination
(B) Enhances chances of self pollination
(C) Reduces chances of cross pollination
(D) Enhances chance of cross pollination

131. Plants with bisexual flowers include
(A) Maize (B) Cocnout
(C) Tulip (D) Custard apple

132. Lesser genetic variability is observed if plant continues
(A) Cross pollination
(B) Self pollination
(C) Meiotic pollination
(D) Mitotic pollination

133. Self pollination is not possible in
(A) Dioecious plants
(B) Maize
(C) Flowers (D) Sunflower

134. Carbohydrates, fats, proteins, vitamins and minerals are form of

(A) Self pollination
(B) Genetic variability
(C) Colonization of plants
(D) Allel variability

135. Bisexual flowers may avoid self pollination if

(A) Either anther or stigma is absent
(B) Anther and stigma do not produce enough eggs and pollens
(C) Anther and stigma mature at different times
(D) Either calyx or corolla is absent

136. In nature, cleistogamous flowers are

(A) Wind pollinated
(B) Insect pollinated
(C) Self pollinated
(D) Bird pollinated

137. Which one of the following is not a correct explanation of cross pollination?

(A) The pollen grains of male flowers are transferred to the stigma of the female
(B) The pollen grains are transferred from one flower to another flower, or flowers of another plant of the same species
(C) The pollen grains are transferred from one flower to another flower situated on same flower
(D) The pollen grains of one flower are transferred to the stigma of the same plant

138. Ornithophily refers to the pollination by which of the following

(A) Snails (B) Birds
(C) Insects (D) Air

139. Plants with ovaries having only one or a few ovules, are generally pollinated by

(A) Birds (B) Butterflies
(C) Bees (D) Wind

140. Plants with ovaries having only one or a few ovules, are generally pollinated by

(A) Birds (B) Butterflies
(C) Bees (D) Wind

141. The pollen are liberated in Cassytha by

(A) Transverse dehiscence
(B) Longitudinal dehiscence
(C) Porous dehiscence
(D) Valvular dehiscence

142. There are 10 flowers in one individual plant of Crotalaria. In each micosporangium of every stamen of all the flowers, there are 30 microspore mother cells. How many pollen grains formed from that plant?

(A) 48,000 (B) 24,000
(C) 10,000 (D) 4,000

143. Both, autogamy and geitonogamy are prevented in

(A) Castor (B) Maize
(C) Papaya (D) Cucumber

144. Even in absence of pollinating agents seed setting is assured in

(A) Salvia
(B) Fig
(C) Commelina
(D) Zostera

145. Wind pollinated flowers are

(A) Small, producing large number of dry pollen grains
(A) Small, brightly coloured, producing large number of pollen grains
(A) Small, producing nectar and dry pollen
(A) Large, producing abundant nectar and pollen

146. Anemophillous flowers have

(A) Sessile stigma
(B) Small smooth stigma
(C) Coloured flower
(D) Large feathery stigma

147. Anthesis is

(A) Opening of flower bud
(B) Pollen mother cell undergoes meiosis
(C) Dehiscence of anthers
(D) Stigma becoming receptive

148. Cleistogamous condition is present in

(A) Brassica (B) *Solanum tuberosum*
(C) Commelina (D) *Allium cepa*

149. Which type of association is found in between entomophilous flower and pollinating agent?

(A) Cooperation (B) Commensalism
(C) Mutualism (D) Coevolution

150. Transfer of pollen grains from the anther to the stigma of another flower of the same plant is called

(A) Geitonogamy (B) Xenogamy
(C) Autogamy (D) Karyogamy

151. In angiosperm, pollen tube liberate their male gametes into the

(A) Egg cell (B) Antipodal cells
(C) Central cell (D) Synergids

152. Insect pollinated flowers are characterized by

(A) Large number of pollens
(B) Dry and smooth pollens
(C) Sticky and rough pollens
(D) Heavy pollens

153. A bisexual flower which never opens up in its life span is known as

(A) Autogamy (B) Allogamy
(C) Homogamy (D) Cleistogamy

154. A hyaline, bisexual and self-fertilized flower that does not open at all is

(A) Chasmogamous
(B) Apogamous
(C) Cleistogamous
(D) Polygamous

155. Allogamy is best favoured by

(A) Dichogamy (B) Dicliny
(C) Cleistogamy (D) Homogamy

156. An interesting modification of flower shape for insect pollination occurs in some orchids in which a male insect mistakes the pattern on the orchid flower for the female of this species and tries to copulate with it, thereby pollinating the flower. This phenomenon is called

(A) Pseudo parthenocarpy
(B) Pseudocopulation
(C) Pseudo pollination
(D) Pseudo apomixis

157. In which one of the following, pollination is autogamous?

(A) Cleistogamy (B) Geitonogamy
(C) Xenogamy (D) Chasmogamy

158. Wind pollination is common in

(A) Orchids (B) Legumes·
(C) Lilies (D) Grasses

159. Fertilization in which male gametes are carried through pollen tube is known as

(A) Porogamy (B) Syngamy
(C) Siphonogamy (D) Chalazogamy

160. Fragrant flowers with well developed nectaries are adapted for

(A) Anemophily (B) Hydrophily
(C) Entomophily (D) Zoophily

161. Germ pore is the region where the exine is

(A) Thin
(B) Uniform
(C) Thin & uniform
(D) Absent

162. A condition in which pollination and fertilization take place in an unopened flower is termed as

(A) Homogamy (B) Apogamy
(C) Polygamy (D) Cleistogamy

163. A generative nucleus in the male gametophyte divides ---------- to form two male gametes

(A) Mitotically
(B) Meiotically
(C) Mitotically and Meiotically
(D) None of these

164. A male gametophyte in angiosperms contains

(A) One gamete only
(B) Two male gamete only
(C) Two male gametes and one tube
(D) One male gamete and one tube nucleus

165. Wind pollination is also known as

(A) Anemophyly (B) Entomophyly
(C) Zoophyly (D) Hydrophyly

166. Barely is a ----------------- pollinated crop

(A) Often-cross (B) Self and cross
(C) Cross (D) Self

167. Cliestogamy encourages

(A) Self pollination
(B) Cross pollination
(C) Apomixes
(D) Inbreeding depression

168. Cotton

(A) Is a self-pollinated crop
(B) Is a cross pollinated crop
(C) Is an often-cross pollinated crop
(D) Is an often-self pollinated crop

169. Development and formation of pollen grains in anther of the stamen is known as

(A) Pollination
(B) Fertilization
(C) Microsporogenesis
(D) Megasporogenesis

170. Development of the pollen tube, the seed and the 3n endosperm, double fertilization is characteristic features of :

(A) Bryophytes (B) Pteridophytes
(C) Gymnosperms (D) Angiosperms

171. Each cell of the female gametophyte is haploid except

(A) Egg cell
(B) Synergied cells
(C) Antipodals cells
(D) Definitive nucleus

172. Each mature anther has

(A) One pollen sac
(B) Two pollen sacs
(C) Three pollen sacs
(D) Four pollen sacs

173. Each megaspore mother cell finally produces

(A) Four megaspores
(B) Two megaspores
(C) Only one megaspore
(D) None of the above

174. Each pollen mother cell produces four haploid cells as a result of

(A) Mitosis
(B) Meiosis
(C) One mitosis and then meiosis
(D) None of the above

175. Embryo-sac is

(A) Male gametophyte
(B) Micro-gametophyte
(C) Saprophyte
(D) Female gametophyte

176. In the flowering plants, the process of fertilization is first discovered by

(A) Robert Brown (B) Strasburger
(C) Virchov (D) Nawaschin

177. In wheat, cross-pollination is

(A) Less than 5% (B) More than 5%
(C) 15% (D) 50%

178. It has been estimated that a maize tassel produces at least ----------- pollen grains for each kernel

(A) 5,000 (B) 15,000
(C) 25,000 (D) 35,000

179. Microsporogenesis occurs is

(A) Root (B) Stem
(C) Anther (D) Ovule

180. Megasporogenesis occurs is

(A) Anther (B) Stamen
(C) Ovary (D) Thalamu

181. Pollination taking place between the flowers borne by the same parent is called

(A) Geitonogamy (B) Alogamy
(C) Xenogamy (D) Syngamy

182. Pollination taking place between two flowers (bisexual or unisexual) borne by two separate plants of the same species is called

(A) Self pollination
(B) Cross pollination
(C) Xenogamy
(D) Geitonogamy

183. Which of the following leads to the cross-pollination?

(A) Unisexuality (B) Dichogamy
(C) Heterostyly (D) Cleistogamy

184. The condition in which the anthers and stigmas of a bisexual flower mature at the same time is known as

(A) Cleistogamy (B) Heterogamy
(C) Syngamy (D) Homogamy

185. The egg on fertilization gives

(A) Secondary
(B) Zygote
(C) Oospore
(D) Primary endospermic nucleus

186. The transference of pollen grain from the anther of a flower to the stigma of the same flower is known as

(A) Self-pollination
(B) Cross pollination
(C) Fertilization
(D) Mating

187. The transference of pollen grains from the anther to the stigma is called

(A) Fertilization
(B) Double fertilization
(C) Mating (D) Pollination

188. Development and formation of pollen grains in anther of the stamen is known as :

(A) Pollination (B) Fertilization
(C) Microsporogenesis
(D) Megasporogenesis

189. Pollination by wind :

(A) Requires abundant pollen production
(B) Required in cross-pollinated crops
(C) Required in self-incompatible lines
(D) All the above

190. Which of the following is incorrect with respect to mode of pollination :

(A) Rice-Wheat-Sorghum-Finger Millet : Self pollinated
(B) Maize-Pearl millet-Oil palm : Cross pollinated
(C) Coconut-Mesta-Cotton : Often cross pollinated
(D) Garlic : No pollination

191. Cross - pollination in maize is mainly on account of which one of the following?

(A) Monoecism
(B) Dioecism
(C) Self - incompatibility
(D) Protogyny

192. ------------ are examples of protandrous species.

(A) Carrots (B) Avocados

(C) Walnuts (D) None of these

193. Leafy vegetables like palak, amaranthus, spinach are :

(A) Self pollinated

(B) Often cross pollinated

(C) Highly cross pollinated

(D) Often self pollinated

194. Pearlmillet is predominantly cross pollinated because it is :

(A) Protogynous (B) Protoandrous

(C) Heterozygous (D) Hermaphrodite

195. Which of the following crops is often cross pollinated?

(A) Ragi (B) Sorghum

(C) Sunflower (D) Maize

ANSWER SHEET

1. **D**	2. **B**	3. **A**	4. **B**	5. **A**	6. **B**	7. **C**	8. **D**	9. **A**	10. **B**
11. **C**	12. **D**	13. **A**	14. **C**	15. **D**	16. **B**	17. **C**	18. **D**	19. **C&D**	20. **C**
21. **D**	22. **A&B**	23. **C**	24. **D**	25. **B**	26. **A**	27. **C**	28. **C**	29. **D**	30. **B**
31. **C**	32. **B**	33. **C**	34. **A**	35. **B**	36. **D**	37. **B**	38. **D**	39. **A**	40. **B**
41. **C**	42. **D**	43. **B**	44. **D**	45. **D**	46. **A**	47. **D**	48. **C**	49. **B**	50. **D**
51. **C**	52. **C**	53. **B**	54. **A**	55. **B**	56. **B**	57. **C**	58. **C**	59. **D**	60. **A**
61. **A**	62. **C**	63. **B**	64. **A**	65. **C**	66. **D**	67. **C**	68. **B**	69. **C**	70. **D**
71. **C**	72. **B**	73. **C**	74. **A**	75. **D**	76. **C**	77. **A**	78. **D**	79. **D**	80. **D**
81. **E**	82. **B**	83. **C**	84. **B**	85. **A**	86. **A**	87. **A**	88. **D**	89. **C**	90. **D**
91. **A**	92. **A**	93. **B**	94. **C**	95. **D**	96. **C**	97. **D**	98. **A**	99. **C**	100. **A**
101. **A**	102. **B**	103. **B**	104. **A**	105. **B**	106. **A**	107. **D**	108. **C**	109. **D**	110. **C**
111. **C**	112. **C**	113. **A**	114. **B**	115. **B**	116. **B**	117. **A**	118. **C**	119. **C**	120. **D**
121. **A&B**	122. **B**	123. **D**	124. **D**	125. **C**	126. **A**	127. **B**	128. **C**	129. **C**	130. **A**
131. **D**	132. **B**	133. **A**	134. **B**	135. **C**	136. **C**	137. **D**	138. **B**	139. **D**	140. **D**
141. **D**	142. **A**	143. **C**	144. **C**	145. **A**	146. **D**	147. **A**	148. **C**	149. **C**	150. **A**
151. **D**	152. **C**	153. **D**	154. **C**	155. **A**	156. **B**	157. **D**	158. **D**	159. **C**	160. **C**
161. **C**	162. **D**	163. **A**	164. **C**	165. **A**	166. **D**	167. **A**	168. **C**	169. **C**	170. **D**
171. **D**	172. **D**	173. **C**	174. **A**	175. **D**	176. **B**	177. **B**	178. **D**	179. **C**	180. **C**
181. **A**	182. **C**	183. **D**	184. **D**	185. **B**	186. **A**	187. **D**	188. **C**	189. **D**	190. **A**
191. **A**	192. **A**	193. **C**	194. **A**	195. **B**					

Chapter 6

FRUIT AND THEIR STRUCTURES

1. Cocoa belongs to which type of fruit?
   (A) Drupe (B) Berry
   (C) Pome (D) Pepo
2. Which one of the following is a drupe?
   (A) Banana (B) Peach
   (C) Papaya (D) Apple
3. Pear is a -------------- fruit
   (A) Drupe (B) Berry
   (C) Pome (D) Pepo
4. Which one of the following is a pome?
   (A) Banana (B) Peach
   (C) Papaya (D) Apple
5. Berry fruits have ------------------ pericarp
   (A) Very hard (B) Hard
   (C) Medium (D) Soft
6. An example for berry is
   (A) Banana (B) Peach
   (C) Apple (D) Mango
7. Which one of the following fruit is a berry?
   (A) Apple (B) Sapota
   (C) Mango (D) Avacado
8. Pumpkin fruit wall is
   (A) Thin (B) Soft
   (C) Thick (D) Papery
9. An example for pepo is
   (A) Apple (B) Sapota
   (C) Mango (D) Cucumber
10. Which one of the following is a hesperidium type of fruit?
    (A) Apple (B) Sapota
    (C) Citrus (D) Papaya
11. An example for siliqua is
    (A) Rice (B) Wheat
    (C) Groundnut (D) Sesamum
12. Which one of the following is a capsule?
    (A) Rice (B) Wheat
    (C) Okra (D) Tomato
13. ------------------ is an example for achene
    (A) Sunflower (B) Castor
    (C) Sesamum (D) Niger
14. Caryopsis type of fruit is present in which crop?
    (A) Groundnut (B) Corn
    (C) Blackgram (D) Greengram
15. A winged achene is
    (A) Nut (B) Achene
    (C) Samara (D) Siliqua
16. Which one of the following is a schizocarp
    (A) Raspberry (B) Strawberry
    (C) Pineapple (D) Carrot
17. ------------ is an aggregate fruit
    (A) Raspberry (B) Pineapple
    (C) Carrot (D) Cabbage

18. Which one of the following is an aggregate fruit?

(A) Mango (B) Citrus
(C) Pineapple (D) Strawberry

19. ---------------- is a multiple fruit?

(A) Mango (B) Citrus
(C) Pineapple (D) Strawberry

20. Fig is a -------------- fruit

(A) Multiple (B) Aggregate
(C) Berry (D) Drupe

21. Sorghum inflorescence is called :

(A) Panicle (B) Flower
(C) Ear (D) Spike

22. Bolting refers to emergence of inflorescence in

(A) Sugarcane (B) Sugarbeet
(C) Potato (D) Sweet potato

23. Inflorescence of wheat plant is called :

(A) Spike (B) Panicle
(C) Arrow (D) Cob

24. Which one of the following is not a fruit?

(A) Tomato (B) Potato
(C) Musk Melon (D) Water melon

25. In flowers, outer whorls are classified as

(A) Haploid whorls
(B) Diploid whorls
(C) Non-reproductive whorls
(D) Reproductive whorls

26. A drupe develops in

(A) Tomato (B) Mango
(C) Wheat (D) Pea

27. Dry indehiscent single-seeded fruit formed from bicarpellary syncarpous inferior ovary is:

(A) Cremocarp (B) Caryopsis
(C) Cypsela (D) Berry

28. Flowers are zygomorphic in

(A) Datura (B) Mustard
(C) Gulmohur (D) Tomato

29. Fruits are mature

(A) Ovules (B) Peduncles
(C) Ovaries (D) Flowers

30. An example of a fruit that disperses seeds by splitting along one or more edges of the carpel is

(A) Samara (B) Follicle
(C) Hesperidium (D) Drupe

31. A fruit developed from hypanthodium inflorescence is called

(A) Caryopsis (B) Hesperidium
(C) Sorosis (D) Syconus

32. Seeds are mature

(A) Ovules (B) Peduncles
(C) Ovaries (D) Flowers

33. A dry, one chambered and one seeded fruit developing from a superior ovary with the hard and woody pericarp is :

(A) Achene (B) Caryopsis
(C) Composite fruit (D) Nut

34. A fruit, which develops from an inflorescence, is known as

(A) Simple fruit
(B) Aggregate fruit
(C) Composite fruit
(D) Drupe fruit

35. A fruit, which is collection of simple fruits developing from the apocarpous pistil of a flower, is known as

(A) Multiple fruit
(B) Aggregate fruit
(C) Composite fruit
(D) Simple fruit

36. A head or capitulum is characteristic of

(A) Compositeae (B) Umbelliferae
(C) Solanaceae (D) Gramineae

37. Fruit of candytuft is known as

(A) Legume (B) Siliqua
(C) Schizocarp (D) Capsule

38. Fruit of pea is known as

(A) Legume (B) Follicle
(C) Siliqua (D) Capsule

39. Fruit of sunflower is

(A) Caryopsis (B) Samara
(C) Achene (D) Nut

40. Fruits of calotropis are

(A) Legume (B) Siliqua
(C) Capsule (D) Follicle

41. Inflorescence of wheat ear is

(A) Catkin (B) Spike
(C) Corymbs (D) Raceme

42. Mango fruit is a

(A) Pepo (B) Pome
(C) Berry (D) Drupe

43. Which of the fruit is boll?

(A) Beans (B) Mustard
(C) Cotton (D) Wheat

44. Orange or lemon fruit is known as

(A) Balusta (B) Pepo
(C) Pome (D) Hesperidium

45. The edible part of the apple is

(A) Mesocarp and endocarp
(B) Endosperm
(C) Only mesocarp
(D) Fleshy thalamus

46. The fleshy edible part of the mango fruit is

(A) Epicarp (B) Endocarp
(C) Mesocarp (D) Pericarp

47. The fruit of Acacia is

(A) Lomentum (B) Double samara
(C) Regma (D) Cypsela

48. The fruit of cumin or coriander is known as

(A) Lomentum (B) Cremocarp
(C) Regma (D) Cypsela

49. The fruits of cucurbittaceae is

(A) Pepo (B) Pome
(C) Berry (D) Drupe

50. The placentation is mustard is

(A) Marginal (B) Parietal
(C) Axile (D) Basal

51. The point of attachment of the body of the ovule is its funicle is known as

(A) Chalaza (B) Raphe
(C) Hilum (D) Stalk

52. The situation where an egg cell is developed into an embryo' without fertilization, is described as

(A) Apomixis
(B) Parthenocarpy
(C) Sexual reproduction
(D) Parthenogenesis

53. The staminate ---------------- plants out yield pistillate plants by about 25 per cent.

(A) Asparagus (B) Lettuce
(C) Spinach (D) Hemp

54. Tomato fruit is a

(A) Drupe (B) Berry
(C) Siliqua (D) Hesperidium

55. When only one fruit develops from the single ovary of a flower it is said to be a

(A) Simple fruit
(B) Aggregate fruit
(C) Multiple fruit
(D) Composite fruit

56. When the ovary is one chambered and the placentae bearing the ovules develop on the inner wall of the ovary, then the placentation is known as

(A) Central (B) Parietal
(C) Axile (D) Free central

57. Fruit of sunflower is :

(A) Caryopsis (B) Cypsela

(C) Nut (D) Samara

58. Botanically pineapple fruit is :

(A) Berry (B) Balusta

(C) Sorosis (D) Pome

59. A fruit which develops from an inflorescence is known as :

(A) Composite fruit

(B) Simple fruit

(C) Aggregate fruit

(D) None of the above

ANSWER SHEET

1. **A**	2. **B**	3. **C**	4. **D**	5. **D**	6. **A**	7. **B**	8. **C**	9. **D**	10. **C**
11. **D**	12. **C**	13. **A**	14. **B**	15. **C**	16. **D**	17. **A**	18. **D**	19. **C**	20. **A**
21. **A**	22. **B**	23. **A**	24. **B**	25. **C**	26. **B**	27. **C**	28. **C**	29. **C**	30. **B**
31. **D**	32. **A**	33. **C**	34. **C**	35. **B**	36. **A**	37. **B**	38. **D**	39. **C**	40. **D**
41. **B**	42. **D**	43. **C**	44. **D**	45. **D**	46. **C**	47. **A**	48. **B**	49. **A**	50. **B**
51. **C**	52. **D**	53. **A**	54. **B**	55. **A**	56. **B**	57. **B**	58. **C**	59. **A**	

Chapter 7

SEED AND ITS STRUCTURE

1. To ensure true to type, the seed must have high --------------- purity
   (A) Physical
   (B) Genetic
   (C) Health
   (D) Both physical and genetic
2. In caryopsis, the seed coat and fruit coat are fused to form
   (A) Pericarp (B) Perisperm
   (C) Tegmen (D) Testa
3. In which crop seed, the aleurone layer is present?
   (A) Redgram (B) Blackgram
   (C) Greengram (D) Barley
4. --------------- is formed from outer integument
   (A) Tegmen (B) Testa
   (C) Seed coat (D) Pericarp
5. Which one of the following is a primary axis?
   (A) Plumule
   (B) Radicle
   (C) Both plumule and radicle
   (D) Both plumule and cotyledon
6. Endosperm is
   (A) 1n (B) 2n
   (C) 3n (D) 4n
7. Embryo is
   (A) 1n (B) 2n
   (C) 3n (D) 4n
8. An example for awned fruit
   (A) Rice (B) Maize
   (C) Pearl millet (D) Sorghum
9. In which crop, the hilum is present?
   (A) Rice (B) Wheat
   (C) Redgram (D) Sesamum
10. In which crop, the caruncle is present?
    (A) Castor (B) Wheat
    (C) Redgram (D) Sesamum
11. Aril is present in
    (A) Castor (B) Nutmeg
    (C) Redgram (D) Sesamum
12. In cotton, --------------- is present
    (A) Awn (B) Caruncle
    (C) Hairs (D) Aril
13. ------------- are present in moringa
    (A) Awn (B) Caruncle
    (C) Hairs (D) Wings
14. A portion of the shoot above the cotyledon is called as
    (A) Epicotyl (B) Hypocotyl
    (C) Coleoptile (D) Coleorhiza
15. What is the name of a portion of the shoot below the cotyledon?
    (A) Epicotyl (B) Hypocotyl
    (C) Coleoptile (D) Coleorhiza

16. Which one of the following protects the plumule?

(A) Epicotyl (B) Hypocotyl
(C) Coleoptile (D) Coleorhiza

17. Which one of the following protects the radicle?

(A) Epicotyl (B) Hypocotyl
(C) Coleoptile (D) Coleorhiza

18. Which one of the following has folded cotyledon?

(A) Safflower (B) Cotton
(C) Sunflower (D) Sesamum

19. Which one of the following has spirally curved embryo?

(A) Tomato (B) Cucumber
(C) Cotton (D) Jute

20. In which one of the dicot seed, a thin layer of endosperm is present?

(A) Safflower (B) Sunflower
(C) Castor (D) Sesamum

21. Cucumber, squashes and beans are

(A) Monocots (B) Dicots
(C) Multicots (D) Ferns

22. Rice, corn and wheat are

(A) Monocots (B) Dicots
(C) Multicots (D) Ferns

23. The aleurone layer in maize grain is specially rich in

(A) Lipids (B) Starch
(C) Proteins (D) Auxins

24. The coconut water and the edible part of coconut are equivalent to

(A) Mesocarp (B) Embryo
(C) Endosperm (D) Endocarp

25. Seed coat of angiosperm seed is also known as

(A) Testa (B) Spores
(C) Hilum (D) Epicotyls

26. A group of flowering plants whose seed typically has two embryonic leaves or cotyledons is called

(A) Dicotyledons
(B) Monocotyledons
(C) Magnolianae
(D) Perispermous

27. An example of a seed with endosperm, perisperm and caruncle is

(A) Castor (B) Cotton
(C) Coffee (D) Lily

28. Cotyledons and testa respectively are edible parts in

(A) Cashew nut and litchi
(B) Groundnut and pomegranate
(C) Walnut and tamarind
(D) French bean and coconut

29. The radicle of corn is enclosed in a

(A) Scutellum (B) Coleorhiza
(C) Hypocotyl (D) Cotyledon

30. Food is stored in seeds in the form of

(A) Starch (B) Minerals
(C) Toxins (D) Fatty acids

31. Which one of the following is not matched correctly?

(A) Plumule-dicot only
(B) Two cotyledons-dicot only
(C) One cotyledon-monocot only
(D) Endosperm-monocot and dicot

32. A seed develops from a (an)

(A) Pollen grain (B) Embryo
(C) Ovule (D) Ovary

33. What would be the number of chromosomes of the aleurone cells of a plant with 42 chromosomes in its root tip cells?

(A) 21 (B) 42
(C) 63 (D) 84

34. Seed coat of angiosperm seed is also known as

(A) Testa (B) Spores

(C) Hilum (D) Epicotyls

35. The larger cells near the micropyle develop into

(A) Ground meristem

(B) Shoot apical meristem

(C) A suspensor

(D) Protoderm

36. Water makes up about ----------------% of a mature seed

(A) One to three

(B) Five to twenty

(C) Twenty five to forty

(D) Forty five to sixty

37. Perisperm develops from a

(A) Embryo sac (B) Ovary

(C) Nucellus (D) Polar nucleii

38. Double fertilization was first discovered by Nawaschin (1898) in

(A) Lilium and Fritillaria

(B) Mango and Sugarcane

(C) Papaya and Pea

(D) Brassica and Candyfruit

39. Egg apparatus consists of

(A) Egg

(B) Egg and polar nuclei

(C) Egg and synergids

(D) Egg and antipodal cells

40. Embryo sac is found in

(A) Endosperm (B) Embryo

(C) Ovule (D) Seed

41. Endosperm is store house for germinating embryo and it contains

(A) Chlorophyll (B) Anthocyanin

(C) MH (D) Fat

42. If a membrane allows the passage of both solvent and solute molecules it is said to be

(A) Non-permeable

(B) Impermeable

(C) Semi-permeable

(D) Permeable

43. In sorghum, somatic chromosome number is 20. What would be its chromosome number in the endosperm?

(A) 10 (B) 20

(C) 30 (D) 40

44. Seeds of monocotyledons are

(A) Albuminous or endospermic

(B) Ex-albuminous or non-endospermic

(C) Cotyledon

(D) Perisperm

45. The effect of the genotype of the pollen grain on the phenotype of the seed is termed as

(A) Apospory

(B) Pollinia

(C) Endopolyploidy

(D) Xenia

46. The fibrous fruit of coconut is dispersed by

(A) Air (B) Water potential

(C) Animals (D) Insects

47. The fleshy food storage tissue, lying immediately within perisperm is called

(A) Endosperm (B) Testa

(C) Caruncle (D) None of the above

48. The seeded fruit of rice and maize is called

(A) Berry (B) Pepo

(C) Siliqua (D) Caryopsis

49. Small size and light weight ensures

(A) Proper seed dispersal

(B) Seed dispersal through wind

(C) Seed dispersal through water

(D) Seed dispersal through insects

50. Plumule refers to the
(A) Developing shoot
(B) Developing root
(C) Developing leaves
(D) Developing cotyledon

51. Some plants have seeds with hooks for
(A) Pollination (B) Fertilization
(C) Dispersion (D) Reproduction

52. Task of fruit for seeds is to
(A) Keep them fresh
(B) Protect them only
(C) Disperse them only
(D) Protect and disperse them

53. Flow of seeds far away from their parent plants is known as
(A) Dispersion (B) Pollination
(C) Fertilization (D) Reproduction

54. The baby plant inside the seed is called the
(A) Ovule (B) Pollen
(C) Embryo (D) Plant

55. A seed is a
(A) Ripened ovule
(B) Ripened ovary
(C) Ripened endosperm
(D) Fused carpel

56. All the following structures are present in dicot seed except
(A) Radicle (B) Endosperm
(C) Coleoptile (D) Seed coat

57. Fruits are mature ---------------- whereas seeds are mature
(A) Ovaries and stamens
(B) Ovules and carpels
(C) Ovules and ovaries
(D) Ovaries and ovules

58. The seed coat is formed from
(A) Integuments (B) Ovary
(C) Endosperm (D) Pericarp

59. A seed contains
(A) Stored food (B) Embryo sac
(C) Placenta (D) Ovule

60. Cotyledons are formed from
(A) Embryos (B) Seeds
(C) Ovaries (D) Ovules

61. All are dicotyledonous seeds except
(A) Mango (B) Castor
(C) Rice (D) Pea

62. Which of the following is a non-endospermic seed
(A) Cereals (B) Millets
(C) Orchids (D) Custard apple

63. The endosperm formed as a part of double fertilization is
(A) Haploid (B) Diploid
(C) Triploid (D) Polyploid

64. Seed coat is derived from
(A) Testa (B) Embryo
(C) Endosperm (D) Nucellus

65. Cotyledons in gymnosperms are called
(A) Embryo
(B) Integuments
(C) Mega-gametophyte
(D) Endosperm

66. Cotyledons in monocots are called
(A) Endosperm
(B) Mega-gametophyte
(C) Embryo
(D) Integuments

67. The coarse rice generally has a -------------------- aleurone layer
(A) Larger
(B) Smaller
(C) Bigger
(D) Smoother

68. Coloured varieties of rice have -------------- aleurone layer

(A) Thinner (B) Thicker
(C) Coarse (D) Smooth

69. In flowering plants, a second seed coat is known as

(A) Integument
(B) Aleurone layer
(C) Tegmen
(D) Inner ventral scale

70. For seed dispersal, plants living on river banks depend on

(A) Water (B) Insects
(C) Wind (D) Animals

71. For seed dispersal, balsam fruits depends upon

(A) Wind
(B) Insects
(C) Explosive mechanism
(D) Water

72. Rubber fruits depend upon seed dispersal mechanism of

(A) Wind
(B) Insects
(C) Explosion
(D) Water

73. Legumes depend upon seed dispersal by

(A) Wind
(B) Insects
(C) Explosive mechanism
(D) Water

74. Peas and beans split along the

(A) Pericarp (B) Sutures
(C) Micropyle (D) Radicle

75. In most species of plants

(A) Radicle (B) Plumule
(C) Cotyledon (D) Zygote

76. Miniature plant contains one or two seed leaves called

(A) Radicle (B) Plomule
(C) Cotyledon (D) Zygote

77. An example of a fruit that disperses seeds by splitting along one or more edges of the carpel is

(A) Samara
(B) Follicle
(C) Hesperidium
(D) Drupe

78. Embryó sac is also known as

(A) Microgametophyte
(B) Megagametophyte
(C) Micosporangium
(D) Megasporangium

79. Number of chromosome in endosperm is

(A) n (B) 2n
(C) 3n (D) 4n

80. Seeds of dicotyledons are

(A) Ex - albuminous or non-endospermic
(B) Albuminous or non-endospermic
(C) Both (A) and (B)
(D) non-endospermic

81. Endosperm is store house for germinating embryo and it contains

(A) Enzymes
(B) Vitamins
(C) Hormones
(D) All the above

82. Gossypol, a toxic substance in cotton is normally present in

(A) Stem (B) Flower
(C) Fruit wall (D) Seed

83. Seeds of pea, gram and beans are

(A) Albuminous
(B) Ex-albuminous
(C) Both A and B
(D) None of the above

84. An example of dicotyledonous albuminous seed is
(A) Castor
(B) Black gram
(C) Sweet pea
(D) Cotton

85. Hard seeds are commonly found in species of
(A) Amaranthaceae
(B) Leguminosae
(C) Alliaceae
(D) Solanaceae

ANSWER SHEET

1. **B**	2. **A**	3. **D**	4. **B**	5. **C.**	6. **C**	7. **B**	8. **A**	9. **C**	10. **A**
11. **B**	12. **C**	13. **D**	14. **A**	15. **B**	16. **C**	17. **D**	18. **B**	19. **A**	20. **C**
21. **B**	22. **A**	23. **C**	24. **C**	25. **A**	26. **A**	27. **A**	28. **B**	29. **B**	30. **A**
31. **A**	32. **C**	33. **C**	34. **A**	35. **C**	36. **B**	37. **C**	38. **A**	39. **C**	40. **B**
41. **D**	42. **D**	43. **C**	44. **A**	45. **D**	46. **B**	47. **A**	48. **D**	49. **B**	50. **A**
51. **C**	52. **D**	53. **A**	54. **C**	55. **A**	56. **B**	57. **D**	58. **A**	59. **A**	60. **A**
61. **C**	62. **C**	63. **C**	64. **A**	65. **C**	66. **A**	67. **A**	68. **B**	69. **C**	70. **A**
71. **C**	72. **C**	73. **C**	74. **B**	75. **C**	76. **C**	77. **B**	78. **B**	79. **C**	80. **A**
81. **D**	82. **D**	83. **B**	84. **A**	85. **B**					

Chapter 8

VEGETATIVE PROPAGATION

1. Which one of the following species is propagated by bulbs?
   (A) Garlic (B) Potato
   (C) Gladiolus (D) Banana
2. Tuber is used to propagate
   (A) Garlic (B) Potato
   (C) Gladiolus (D) Banana
3. Gladiolus is propagated by
   (A) Bulb (B) Tuber
   (C) Corm (D) Sucker
4. Banana is propagated by
   (A) Bulb (B) Tuber
   (C) Corm (D) Sucker
5. Potato is propagated by
   (A) Bulb (B) Tuber
   (C) Corm (D) Sucker
6. Which one of the following is propagated by rhizome
   (A) Garlic (B) Potato
   (C) Turmeric (D) Gladiolus
7. Ginger is propagated by
   (A) Tuber (B) Rhizome
   (C) Runner (D) Stolon
8. Strawberry is propagated by
   (A) Tuber (B) Rhizome
   (C) Runner (D) Stolon
9. Which one of the following crop is propagated by stolon?
   (A) Potato (B) Cardamom
   (C) Strawberry (D) Grass
10. --------------- is propagated by offshoot
   (A) Potato (B) Cardamom
   (C) Pineapple (D) Grass
11. Which one of the following crop is propagated by crown?
   (A) Potato (B) Cardomam
   (C) Pineapple (D) Turmeric
12. --------------- is propagated by bulbil
   (A) Potato (B) Begonia
   (C) Chlorophytum (D) Saccharum
13. Which one of the following crop is propagated by bulbil
   (A) Begonia (B) Potato
   (C) Chlorophytum (D) Saccharum
14. Date palm is propagated by
   (A) Corm (B) Bulb
   (C) Offshoot (D) Runner
15. How ferns are being propagated?
   (A) Corm (B) Bulb
   (C) Offshoot (D) Runner
16. Cork cambium is
   (A) Promeristem
   (B) Ground meristem
   (C) Lateral meristem
   (D) Intercalary meristem

17. In agave, vegetative propagation is carried out through

(A) Offshoots (B) Rhizome
(C) Stolon (D) Sucker

18. Which is the example of root crop

(A) Potato (B) Sweet potato
(C) Groundnut (D) Carrot

19. Which method of plant propagation involves the use of girdling?

(A) Grafting (B) Cuttings
(C) Layering (D) Micropropagation

20. Consider the following statements and choose the correct option.

(A) The genetic constitution of a plant is unaffected in vegetative propagation
(B) Rhizome in ginger serves as an organ of vegetative reproduction
(C) Totipotency of cells enables us to micropropagate plants

(A) Statement (B) alone is true
(B) Statements (A) and (B) alone are true
(C) Statements (B) and (C) alone are true
(D) All the three statements (A), (B) and (C) are true

21. Grafting is successful in dicots, but not in monocots because the dicots have

(A) Cork cambium
(B) Vessels with elements arranged end to end
(C) Vascular bundles arranged in a ring
(D) Cambium for secondary growth

22. In which one pair both the plants can be vegetatively propagated by leaf pieces?

(A) Asparagus and Bryophyllum
(B) Agave and Kalanchoe
(C) Bryophyllum and Kalanchoe
(D) Chrysanthemum and Agave

23. A clone is a group of individuals obtained through

(A) Self pollination
(B) Vegetative propagation
(C) Hybridization
(D) Cross pollination

24. Axillary buds are used to raise crop of

(A) Wheat (B) Rice
(C) Groundnut (D) Sugarcane

25. Banana plant develops from

(A) Rhizome (B) Seed
(C) Sucker (D) Stolon

26. A corm is

(A) A swollen rot
(B) A compressed, swollen, vertical underground stem
(C) Swollen horizontal underground stem
(D) The swollen end of an underground stem

27. Clinging roots are found in

(A) Orchids (B) Trapa
(C) Podostemon (D) Screwpine

28. Fleshy adventitious roots are found in

(A) Potato and onion
(B) Radish and carrot
(C) Dahlia and Sweet potato
(D) Turnip and rice

29. Ginger multiplies vegetatively by

(A) Bud (B) Stem
(C) Tuber (D) Rhizome

30. In Agave, vegetative propagation is carried out through

(A) Bulbils (B) Rhizome
(C) Stolon (D) Sucker

31. In Bryophyllum, buds arise from the notches on the margin of the

(A) Roots (B) Stem
(C) Leaves (D) Nodes

32. Buds drop off from the leaf and grow into new plants in

(A) Bryophyllum (B) Potato
(C) Onion (D) Ginger

33. Which of the following are ornamental and fruit yielding plants

(A) Mangoes and Lemon
(B) Crotans
(C) Sapota and Oranges
(D) All the above

34. Vegetative propagation takes place in

(A) Sugarcane (B) Rice
(C) Sorghum (D) Maize

35. Propagation by stem cutting takes place in

(A) Tomato (B) Begonia
(C) Rose (D) Bryophyllum

36. A piece of ginger develops into a new plant if it is planted in the soil along with the

(A) Leaf (B) Bud
(C) Stem (D) Seeds

37. Finger like projections in ginger are called

(A) Scale leaf (B) Node
(C) Buds (D) Notches

38. Which of the following plant grow by spreading on the ground?

(A) Grass (B) Onion
(C) Potato (D) Colacasia

39. Vegetative propagation in plantain takes place by

(A) Nodes (B) Buds
(C) Tuberous stem (D) Roots

40. Buds sprout from tuberous stem are called

(A) Sword suckers (B) Stigma
(C) Thalamus (D) Anther

41. Propagation by layering takes place in

(A) Jasmine (B) Marigold
(C) Crossandra (D) Petunia

42. In grafting, the plant that is rooted in the soil is called

(A) Stock (B) Stamen
(C) Scion (D) Graft

43. In grafting, the cutting that is attached to the stock is called

(A) Corolla (B) Calyx
(C) Scion (D) Thalamus

44. Which of the following is selected grafting to get desired qualities?

(A) Stigma (B) Scion
(C) Stock (D) Style

45. Which of the following is supplied by the stock to scion?

(A) Water
(B) Organic substances
(C) Air only
(D) Oxygen only

46. Which of the following statement is true regarding vegetative propagation?

(A) Vegetative propagation occurs naturally and artificially producing genetically identical plants
(B) Vegetative propagation occurs naturally and artificially producing genetically different plants
(C) Vegetative propagation occurs artificially producing genetically similar plants
(D) Vegetative propagation occurs naturally producing genetically similar plants

47. All the following plants can be propagated by leaf cuttings except

(A) Begonia (B) Hibiscus
(C) Bryophyllum (D) Sansiveria

48. Which of the following method is commonly practiced for root induction?

(A) Cutting (B) Layering
(C) Grafting (D) Budding

49. Rooting in stem cuttings are stimulated by using

(A) Jasmonic acid (B) ABA
(C) Ethylene (D) IAA

50. 'Conifer cuttings' are practiced for propagating pine, spruce etc. They are usually taken during

(A) Early spring (B) Early winter
(C) Late spring (D) Late winter

51. The method of inducing rooting by girdling aerial branches is called air layering. Air layering is also called as

(A) Root layering
(B) Chinese layering
(C) Soil layering
(D) Grafting

52. The method of layering in which rooting is induced at different nodes simultaneously in the soil is called compound layering. It is also called as

(A) Air layering
(B) Serpentine layering
(C) Trench layering
(D) Mound layering

53. Which of the following method is suitable for combining the desirable characters of two plants together in a single plant

(A) Cutting (B) Layering
(C) Grafting (D) Girdling

54. In grafting, the rooted plant is called the stock. The stem cutting from the donor plant is called

(A) Transplant (B) Scion
(C) Bud graft (D) Root stock

55. In vegetative propagation, which of the following parts produce new plants?

(A) Fruit (B) Seed
(C) Branch (D) Leaves

56. Vegetative propagation through leaves takes place in

(A) Bryophyllum (B) Ginger
(C) Potato (D) Tapioca

57. Vegetative propagation through roots takes place in

(A) Curry leaf (B) Sweet potato
(C) Jasmine (D) Bryophyllum

58. In some plants like sweet potato and tapioca food is stored in roots. They are called

(A) Tuberous roots (B) Notches
(C) Buds (D) Nodes

59. Vegetative propagation by stems takes place in

(A) Marigold (B) Onion
(C) Carrot (D) Jasmine

60. Radish is an example of

(A) Fusiform root (B) Napiform root
(C) Conical root (D) Tuberous root

61. Zamikand and Taro (Colocasia) are example of

(A) Rhizome (B) Tuber
(C) Bulb (D) Corm

62. A method of breeding most commonly used when the desired variation is required to be induced in a vegetatively propagated crop is

(A) Polyploidy
(B) Pedigree method of breeding
(C) Mutation breeding
(D) Backcross method of breeding

63. Gibberellins are now commercially used to increase sugarcane growth and sugar yields at

(A) China (B) Hawaii
(C) Japan (D) Korea

64. Ginger and turmeric are example of modified stem

(A) Rhizome (B) Tuber
(C) Bulb (D) Corm

65. In asparagus one or more short, green, cylindrical or sometimes flattened branches of limited growth develop from node of stem, such a branch is known as

(A) Cladode (B) Phylloclade
(C) Stem tendril (D) Thorn

66. In the pitcher plant, the pitcher is modification of

(A) Stem (B) Root
(C) Leaf (D) Stem-branch

67. Onion and garlic are example of

(A) Rhizome (B) Tuber
(C) Bulb (D) Corm

68. Phylloclade is

(A) Underground modification of stem
(B) Sub-aerial modification of stem
(C) Aerial modification of stem
(D) Root modification

69. Potato and Jerusalem artichoke are example of

(A) Rhizome (B) Tuber
(C) Bulb (D) Corm

70. Which of the following is sub-aerial modification of stem?

(A) Corm (B) Stolon
(C) Thorn (D) Bulb

71. The botanical name of the onion is

(A) *Allium sativum*
(B) *Allium cepa*
(C) *Pisum sativum*
(D) *Tagetes erecta*

72. Plants that attach themselves to any neighbouring object often by means of some special devices and climb it are called

(A) Trailers (B) Creepers
(C) Climbers (D) None of these

73. Stilt roots are produced from

(A) Main root
(B) Stem and branches
(C) Secondary roots
(D) Main stem

ANSWER SHEET

1. **A**	2. **B**	3. **C**	4. **D**	5. **B**	6. **C**	7. **B**	8. **C**	9. **D**	10. **C**
11. **C**	12. **A**	13. **C**	14. **C**	15. **C**	16. **C**	17. **A**	18. **D**	19. **C**	20. **D**
21. **D**	22. **C**	23. **B**	24. **D**	25. **C**	26. **B**	27. **B**	28. **B**	29. **D**	30. **A**
31. **C**	32. **A**	33. **D**	34. **A**	35. **C**	36. **B**	37. **C**	38. **A**	39. **C**	40. **A**
41. **A**	42. **A**	43. **C**	44. **B**	45. **A**	46. **A**	47. **B**	48. **B**	49. **D**	50. **B**
51. **B**	52. **B**	53. **C**	54. **B**	55. **D**	56. **A**	57. **B**	58. **A**	59. **D**	60. **A**
61. **D**	62. **A**	63. **B**	64. **A**	65. **A**	66. **C**	67. **C**	68. **C**	69. **B**	70. **B**
71. **B**	72. **B**	73. **B**							

Chapter 9

APOMIXIS

1. Development of seed without fertilization is known as
   (A) Sexual reproduction
   (B) Asexual reproduction
   (C) Apomixis
   (D) Polyembryony

2. Which one of the following crop is an example for haploid parthenogenesis?
   (A) Cucurbits (B) Solanum
   (C) Hibiscus (D) Gossypium

3. The development of the embryo from some other cell of embryo sac other than egg is defined as
   (A) Apomixis (B) Parthenocarpy
   (C) Polyembryony (D) Apogamy

4. An example for apogamy is
   (A) Alluim (B) Solanum
   (C) Cucurbits (D) Gossypium

5. The development of embryo sac from the cell of the archesporium is known as
   (A) Generative apospory
   (B) Somatic apospory
   (C) Diploid parthenogenesis
   (D) Diploid apogamy

6. Which one of the following crop is an example for adventives embryony?
   (A) Mango (B) Sapota
   (C) Citrus (D) Banana

7. A fruit is best defined as
   (A) Ripened ovule enclosing one or more seeds
   (B) Ripened ovary enclosing one or more seeds
   (C) Ripened endosperm enclosing one or more seeds
   (D) Ripened nucellus enclosing one or more seeds

8. A true fruit is the one where
   (A) Only ovary of the flower develops into fruit
   (B) Ovary and calyx of the flower develops into fruit
   (C) Ovary, calyx and thalamus of the flower develops into fruit
   (D) All floral whorls of the flower develops into fruit

9. Arrange the following in correct order
   1. Apple -- A) Drupe
   2. Pea -- B) Hespiridium
   3. Orange -- C) Legume
   4. Mango -- D) Pome
   (A) 1-A, 2-C, 3-D, 4-B
   (B) 1-D, 2-B, 3-A. 4-C
   (C) 1-D, 2-C, 3-B, 4-A
   (D) 1-B, 2-C, 3-A, 4-C

10. The edible fleshy part of mango is called
    (A) Pericarp (B) Mesocarp
    (C) Endocarp (D) Epicarp

11. Ovary wall gives rise to
(A) Fruit wall (B) Seed coat
(C) Mesocarp (D) Endocarp

12. Development of fruits without fertilization is called
(A) Apospory (B) Parthenogenesis
(C) Parthenocarpy (D) Polygamy

13. The water of coconut is
(A) Liquid mesocarp
(B) Liquid endocarp
(C) Degenerated liquid endosperm
(D) Liquid nucellus

14. The fruit of rice and wheat is called
(A) Achene (B) Siliqua
(C) Follicle (D) Caryopsis

15. In tomato, fruit is a
(A) Drupe (B) Berry
(C) Pepo (D) Pome

16. Arrange the following in correct order
1. Lometum --- A) Fig
2. Regma -- B) Groundnut
3. Sorosis -- C) Castor
4. Syconus -- D) Pine apple
(A) 1-A, 2-C, 3-D, 4-B
(B) 1-B, 2-D, 3-A, 4-C
(C) 1-C, 2-A, 3-D, 4-B
(D) 1-B, 2-C, 3-D, 4-A

17. Apomictic embryos in citrus arise from
(A) Material sporophytic tissue in ovule
(B) Syneroids
(C) Diploid egg
(D) Antipodal cells

18. In a type of apomixis known as adventives embryony, embryos develop directly from the
(A) Zygote
(B) Nucellus or integuments
(C) Synergids or antipodals in an embryo sac
(D) Accessory embryo sacs in the ovule

19. Apomixis is a type of reproduction that results in the development of a/an
(A) New organism without fusion of gametes
(B) New organism from fusion products of gametes
(C) Embryo from endosperm
(D) Embryo from nucellus

20. Nucellar polyembryony is reported in species of
(A) Brassica (B) Citrus
(C) Gossypium (D) Triticum

21. If any somatic cell of sporophyte produces gametophyte without reduction division, it is
(A) Parthenogenesis
(B) Apogamy
(C) Apospory
(D) Amphimixis

22. Development of a fruit without fertilization is known as
(A) Parthenogenesis
(B) Polyembryo
(C) Budding
(D) Apomixes

23. Development of seed without sexual process is
(A) Apomixes (B) Polyembryony
(C) Parthenogenesis
(D) Polygamous

24. The situation where an egg cell is developed into an embryo' without fertilization, is described as
(A) Apomixis
(B) Parthenocarpy
(C) Sexual reproduction
(D) Parthenogenesis

25. A type of apomixis in which embryo sac develops from vegetative cells of ovule :

(A) Apospory

(B) Apogamy

(C) Diplospory

(D) Polyembryony

26. In apomixes, the progeny closely resembles :

(A) Male parent

(B) Female parent

(C) Both male and female parents

(D) Restorer parent

ANSWER SHEET

1. **C**	2. **B**	3. **D**	4. **A**	5. **A**	6. **C**	7. **B**	8. **A**	9. **C**	10. **B**
11. **A**	12. **C**	13. **D**	14. **D**	15. **B**	16. **D**	17. **A**	18. **B**	19. **A**	20. **B**
21. **C**	22. **A**	23. **A**	24. **D**	25. **C**	26. **D**				

Chapter 10

SYNTHETIC SEED

1. Who has patented the synthetic seed

   (A) Murashige (B) Skoog

   (C) Martin (D) Redenbaugh

2. In desiccated system of synthetic seed, which one of the following chemical is used for coating somatic embryos?

   (A) PEG

   (B) Hydrogel

   (C) Sodium alginate

   (D) Calcium alginate

3. In hydrated system of synthetic seed, which one of the following chemical is used for coating somatic embryos?

   (A) PEG

   (B) Sodium alginate

   (C) Dolomite

   (D) Activated clay

4. A technique of micropropagation is

   (A) Multiple root production

   (B) Somatic embryogenesis

   (C) Growth of microorganisms on culture

   (D) Multiple shoot production and embryo

5. Synthetic seeds are

   (A) Naturally synthesized seeds

   (B) Somatic embryos encapsulated in suitable matrix

   (C) Seeds of plants modified genetically

   (D) None of these

6. Somatic embryoids are

   (A) Identical with zygotic embryos and without seed coats

   (B) Identical with zygotic embryos and with seed coats

   (C) Non-identical with zygotic embryos and without seed coats

   (D) Non-identical with zygotic embryos and with seed coats

7. The production of high quality and uniform embryos has been limited to only

   (A) Carrot

   (B) Alfalfa

   (C) Both (A) and (B)

   (D) Sandalwood

8. The preserved embryoids are termed as

   (A) Synthetic seeds

   (B) Semi-synthetic seeds

   (C) Natural seeds

   (D) Fermented seeds

9. Encapsulation is necessary to produce and protect synthetic seeds. The encapsulation is carried out by various types of hydrogels, which are

   (A) Soluble in water

   (B) Soluble in organic solvents

   (C) Insoluble in water

   (D) Insoluble in organic solvents

10. Which of the following is not true about synthetic seeds?

(A) Cab be stored for a year without the loss of valuables

(B) Easy to handle

(C) Can be directly sown in the soil like natural seeds

(D) Need hardening in the green house

11. Recalcitrant seeds are

(A) Resistant to drying and freezing temperatures

(B) Killed by drying and freezing temperatures

(C) Both (A) and (B)

(D) None of the above

ANSWER SHEET

1. **D**	2. **A**	3. **B**	4. **B**	5. **B**	6. **C**	7. **A**	8. **A**	9. **D**	10. **B**
11. **B**									

Chapter 11

SEED FORMATION AND DEVELOPMENT

1. Developing ovule is commonly attached to the placenta by the
   (A) Hilum (B) Chalaza
   (C) Raphe (D) Funicle
2. What is the part of a seed opposite to micropyle?
   (A) Hilum (B) Chalaza
   (C) Raphe (D) Funicle
3. What is the part of a seed between the chalaza and hilum?
   (A) Hilum (B) Chalaza
   (C) Raphe (D) Funicle
4. The scar on the ovule made where the funiculus detatches at maturity is known as
   (A) Hilum (B) Chalaza
   (C) Raphe (D) Funicle
5. Which one of the following part of a seed is an outgrowth of integument?
   (A) Aril (B) Chalaza
   (C) Raphe (D) Funicle
6. Which one of the following is an outgrowth of integument?
   (A) Hilum (B) Caruncle
   (C) Raphe (D) Funicle
7. Which one of the following is an appendage?
   (A) Hilum (B) Chalaza
   (C) Strophiole (D) Funicle
8. From which part of the ovary, the achesporial cell develops?
   (A) Nucleus (B) Nucellus
   (C) Ovule (D) Integument
9. How many cell division takes place to produce megaspores from megaspore mother cell.
   (A) One-step meiosis
   (B) Two-step meiosis
   (C) Three-step meiosis
   (D) Four-step meiosis
10. What is the arrangement of cells in the embryosac?
    (A) 3 cells at one end + 3 cells at centre + 2 cells at other end
    (B) 2 cells at one end + 2 cells at centre + 4 cells at other end
    (C) 3 cells at one end + 2 cells at centre + 3 cells at other end
    (D) 2 cells at one end + 3 cells at centre + 3 cells at other end
11. In which stage of the embryogeny, the embryo develops?
    (A) Globular stage
    (B) Heart stage
    (C) Torpedo stage
    (D) Mature embryo

12. In which stage of the embryogeny, the suspensor develops?
(A) Globular stage
(B) Heart stage
(C) Torpedo stage
(D) Mature embryo

13. The nuclear compliment of endosperm is
(A) Haploid (B) Diploid
(C) Triploid (D) Tetraploid

14. The nuclear compliment of embryo is
(A) Haploid (B) Diploid
(C) Triploid (D) Tetraploid

15. The zygote develops into
(A) Pedicel (B) Embryo
(C) Stigma (D) Style

16. One of the male nuclei fuses with the egg cell and forms
(A) Stigma (B) Pedicel
(C) Corolla (D) Zygote

17. One of the nucleus of the pollen tube fuses with the secondary nucleus of the ovum and grows into
(A) Stamen (B) Stigma
(C) Anther (D) Endosperm

18. Endosperm formation begins with
(A) The establishment of the suspensor
(B) The fusion of the antipodals
(C) The fertilization of the polar nuclei
(D) The syncytial development of the embryo

19. The establishment of the shoot-root axis begins with the
(A) Formation of the nucleus
(B) Radial division of the embryo
(C) Location of the micropyle
(D) Formation of the suspensor

20. The seed coat of angiosperm forms from the
(A) Ovary (B) Carpel
(C) Ovule (D) Peduncle

21. Number of chromosome in endosperm is
(A) 2n (B) 3n
(C) n (D) 4n

22. Sporogenesis means
(A) Formation of zygote
(B) Formation of embryo
(C) Germination of seed
(D) Development and formation of spores

23. The fusion of a male gamete with two polar nuclei is often termed
(A) Fertilization
(B) Syngamy
(C) Triple fusion
(D) Pollination

24. The life cycle of an angiosperm is consist of
(A) Sporophytic stage
(B) Ovule stage
(C) Ovary stages
(D) Placentation stage

25. The main type of endosperm development is
(A) Nuclear type
(B) Perisperm type
(C) Non nuclear type
(D) Nucellus type

26. The nucleus of the functional megaspore divides
(A) Meiotically
(B) Asexually
(C) Apomictically
(D) Mitotically

27. The nutritive tissue, which is infact, a remnant of the nucleus is called
(A) Endosperm (B) Perisperm
(C) Yolk (D) None of the above

28. The phenomenon is called mesogamy, when the tube enters into the embryo sac through

(A) Micropyle (B) Chalaza
(C) Integuments (D) None of the above

29. The pollen along with pollen tube is known as

(A) Microspore (B) Pollengrain
(C) Megaspore (D) Male gametophyte

30. The process of megasporogenesis occurs in

(A) Embryosac (B) Egg cell
(C) Pollensac (D) Ovule

31. The process, double fertilization was first discovered by

(A) Nawaschin (B) Strasburger
(C) Robert Brown (D) Hugo de Vries

32. The production of sperms is known as

(A) Megagametogenesis
(B) Microsporogenesis
(C) Microgametogenesis
(D) Megasporogenesis

33. The two integuments of a ovule develop into

(A) Seed coats (B) Disintegrate
(C) Plumule (D) Pericarp

34. When the male gametes are shed into embryo sac through microphyle is called

(A) Porogamy (B) Chalazogamy
(C) Mesogamy (D) Syngamy

35. When the seed coat forms,the metabolic activity

(A) Slows down (B) Accelerates
(C) Fluctuates (D) Ceases

36. At the end opposite the micropyle, the first meristem to differentiate in gymnosperm embryos is the

(A) Ground meristem
(B) Shoot apical meristem
(C) Root apical meristem
(D) Protoderm

37. In the angiosperm embryo, the cells near the suspensor are destined to form

(A) Cotyledon (B) Root
(C) Endosperm (D) Plumule

38. The outermost cells in a plant embryo become

(A) Vascular tissue
(B) Ovule
(C) Ground tissue
(D) Epidermal tissue

39. The first cell division of the plant zygote is typically

(A) Longitudinal (B) Diagonal
(C) Transverse (D) Radial

40. As the spherical embryo develops, all of the tissues listed below develop except

(A) Ground meristem
(B) Proembryo
(C) Procambium
(D) Apical meristem

41. The suspensor formed in angiosperm seed development links

(A) Roots to shoots
(B) Stem to leaves
(C) Embryo to nucellus
(D) Cotyledons to epicotyl

42. In plant embryo development, the bulk of the embryonic interior is

(A) Epidermal cells
(B) Ground tissue
(C) Mesodermal tissue
(D) Vascular tissue

43. In angiosperm embryos, the shoot apical meristem may be located at the tips of the

(A) Hypocotyl (B) Plumule
(C) Radicle (D) Epicotyl

44. Embryo is called monosporic when it develops from

(A) All the four megaspores
(B) Only from two functional megaspores
(C) Three megaspores
(D) One of the megaspores out of the four megaspores which are derived from division of megaspore mother cell

45. Nucellar polyembryony is reported in the species of

(A) Brassica (B) Citrus
(C) Gossypium (D) Triticum

46. Endosperm of angiosperm results after fertilization from

(A) Synergids
(B) Secondary nucleus
(C) Antipodal cells
(D) Egg

47. Filiform apparatus is found in

(A) Synergids
(B) Anther wall
(C) Secondary nucleus
(D) Egg cell

48. Fertilization in which male gametes are carried through pollen tube is known as

(A) Porogamy (B) Syngamy
(C) Siphonogamy (D) Chalazogamy

49. Fusion of two dissimilar gametes is known as

(A) Fertilization (B) Autogamy
(C) Allogamy (D) Dichogamy

50. Germ pore is the region where the exine is

(A) Thin
(B) Uniform
(C) Thick and Uniform
(D) Absent

51. How many microspore mother cell will produce 1000 micropores?

(A) 100 (B) 200
(C) 250 (D) 500

52. If any somatic cell of sporophyte produces gametophyte without reduction division, it is

(A) Parthenogenesis
(B) Apogamy
(C) Apospory
(D) Amphimixis

53. Development of embryo either from synergids or antipodal cell is referred to as

(A) Apospory
(B) Adventive embryony
(C) Apogamy
(D) Diplospory

ANSWER SHEET

1. **D**	2. **B**	3. **C**	4. **A**	5. **A**	6. **B**	7. **C**	8. **B**	9. **B**	10. **C**
11. **A**	12. **A**	13. **C**	14. **B**	15. **B**	16. **D**	17. **D**	18. **C**	19. **D**	20. **C**
21. **B**	22. **D**	23. **C**	24. **A**	25. **A**	26. **D**	27. **B**	28. **C**	29. **D**	30. **D**
31. **A**	32. **C**	33. **A**	34. **A**	35. **D**	36. **C**	37. **B**	38. **D**	39. **C**	40. **B**
41. **C**	42. **B**	43. **D**	44. **D**	45. **B**	46. **B**	47. **A**	48. **C**	49. **A**	50. **D**
51. **C**	52. **C**	53. **C**							

Chapter 12

SEED MATURITY

1. Harvesting of seed should be done at physiological maturity to avoid harvesting of
   (A) Immature and filled seed
   (B) Mature and ill filled seed
   (C) Mature and filled seed
   (D) Immature and ill filled seed

2. What is the moisture content of seed at the time of physiological maturity?
   (A) 25-30% (B) 20-25%
   (C) 30-35% (D) 35-40%

3. Which one of the following decides the physiological maturity?
   (A) Minimum dry weight and seed germination
   (B) Maximum dry weight and seed germination
   (C) Minimum dry weight and maximum seed germination
   (D) Maximum dry weight and minimum seed germination

4. What is the seed moisture at harvestable maturity?
   (A) 10-14% (B) 14-16%
   (C) 16-18% (D) 18-20%

5. Days to physiological maturity for rice is
   (A) 21-24 days
   (B) 25-28 days
   (C) 28-31 days
   (D) 32-35 days

6. Choose the correct visual symptom from the following to fix the stage of physiological maturity in rice
   (A) Formation of black spot on seed coat
   (B) Formation of black spot on endosperm
   (C) Formation of black spot on scutellum
   (D) Formation of black spot on aleurone layer

7. Days to physiological maturity for soybean is ------ days after anthesis
   (A) 35 (B) 40
   (C) 45 (D) 50

8. How many days sunflower seed take to reach physiological maturity?
   (A) 30-36 days after flowering
   (B) 32-38 days after flowering
   (C) 34-40 days after flowering
   (D) 36-42 days after flowering

9. Mention the number of days taken to reach physiological maturity in cotton
   (A) 50 days after anthesis with boll cracking
   (B) 52 days after anthesis with boll cracking
   (C) 54 days after anthesis with boll cracking
   (D) 56 days after anthesis with boll cracking

10. Niger seed reaches its physiological maturity at -------------- days after sowing

(A) 25 (B) 30
(C) 35 (D) 40

11. What is the visual symptom for physiological maturity of ashgourd?

(A) Formation of white coat on skin
(B) Formation of powdery coat on skin
(C) Formation of ashy coat on skin
(D) Formation of grey coat on skin

12. What is the optimum stage of harvesting of cauliflower seed

(A) 7 weeks after anthesis
(B) 6 weeks after anthesis
(C) 5 weeks after anthesis
(D) 4 weeks after anthesis

13. Days taken for physiological maturity of papaya seed is ----------- days after anthesis

(A) 120-127 (B) 126-133
(C) 132-139 (D) 138-145

14. In ocimum species, the seeds are harvested at -------------- days after anthesis.

(A) 18-30 (B) 16-30
(C) 14-30 (D) 12-30

15. Prosopis seed matures at ------------------ days after anthesis

(A) 88 (B) 89
(C) 90 (D) 91

ANSWER SHEET

1. **D**	2. **A**	3. **B**	4. **D**	5. **C**	6. **D**	7. **C**	8. **A**	9. **B**	10. **C**
11. **C**	12. **A**	13. **B**	14. **A**	15. **D**					

Chapter - 13

CHEMICAL COMPOSITION OF SEEDS

1. What is the percentage of amylose in starch granules?
   (A) 20-30% (B) 25-30%
   (C) 15-30% (D) 10-30%

2. What is the percentage of amylopectin in starch granules?
   (A) 40-80% (B) 50-80%
   (C) 60-80% (D) 70-80%

3. Where the food reserves are stored in wheat?
   (A) Megagametophyte
   (B) Endosperm
   (C) Cotyledons
   (D) Perisperm

4. The food reserves are stored in ------------- in the case of castor bean
   (A) Megagametophyte
   (B) Endosperm
   (C) Cotyledons
   (D) Perisperm

5. In pine, ----------- is a food storage tissue
   (A) Megagametophyte
   (B) Endosperm
   (C) Cotyledons
   (D) Perisperm

6. Which one of the following organ is a food storage tissue of peanut?
   (A) Megagametophyte
   (B) Endosperm
   (C) Cotyledons
   (D) Perisperm

7. Starch is a polymer of
   (A) Fructose (B) Glucose
   (C) Dextrose (D) Galactose

8. Amylose is a straight chain polymer of glucose units joined by ------------ linkages
   (A) Alpha 1-6 glycosidic
   (B) Beta 1-6 glycosidic
   (C) Alpha 1-4 glycosidic
   (D) Beta 1-4 glycosidic

9. Amylopectin has -------------- bonds
   (A) Alpha 1-6 glycosidic
   (B) Beta 1-6 glycosidic
   (C) Alpha 1-4 glycosidic
   (D) Beta 1-4 glycosidic

10. Which one of the following has more amylopectin than amylose?
    (A) Gluclose (B) Fructose
    (C) Dextrose (D) Starch

11. Cellulose is a polymer of starch with ------- linkages
    (A) Alpha 1-4 (B) Beta 1-4
    (C) Alpha 1-6 (D) Beta 1-6

12. Which one of the following is a monosaccharide

(A) Starch (B) Sucrose
(C) Glucose (D) Cellulose

13. Which one of the following is a disaccharide?

(A) Starch (B) Sucrose
(C) Glucose (D) Cellulose

14. Which one of the following is a polysaccharide?

(A) Starch (B) Sucrose
(C) Glucose (D) Fructose

15. In galactomannan, glucose and mannose are joined by ----------------- linkages

(A) α 1-4 (B) α 1-6
(C) ß 1-4 (D) ß 1-6

16. In which of the following crop seed, the albumin is predominant

(A) Sunflower (B) Oats
(C) Sorghum (D) Bread wheat

17. In which of the following crop seed, the globulin is predominant

(A) Sunflower (B) Oats
(C) Sorghum (D) Bread wheat

18. In which of the following crop seed, the prolamin is predominant

(A) Sunflower (B) Oats
(C) Sorghum (D) Bread wheat

19. In which of the following crop seed, the glutelin is predominant

(A) Sunflower (B) Oats
(C) Sorghum (D) Bread wheat

20. In rapeseed, --------- storage protein is predominant

(A) Glutelin (B) Prolamin
(C) Globulin (D) Albumin

21. The storage protein in rice is

(A) Glutelin (B) Prolamin
(C) Globulin (D) Albumin

22. ------------- is predominant in maize

(A) Glutelin (B) Prolamin
(C) Globulin (D) Albumin

23. Wheat has more

(A) Glutelin (B) Prolamin
(C) Globulin (D) Albumin

24. Which one of the following storage protein is soluble in water?

(A) Albumin (B) Globulin
(C) Prolamin (D) Glutelin

25. Which one of the following storage protein is soluble in salt solution?

(A) Albumin (B) Globulin
(C) Prolamin (D) Glutelin

26. Which one of the following storage protein is soluble in alcohol?

(A) Albumin (B) Globulin
(C) Prolamin (D) Glutelin

27. Which one of the following storage protein is soluble in dilute acids or bases?

(A) Albumin (B) Globulin
(C) Prolamin (D) Glutelin

28. Which one of the following becomes solid at room temperature?

(A) Fat (B) Oil
(C) Lipid (D) Triglyceride

29. ------------- becomes liquid at room temperature

(A) Fat (B) Oil
(C) Lipid (D) Triglyceride

30. -------------- becomes liquid and solid at room temperature

(A) Fat (B) Oil
(C) Lipid (D) Triglyceride

31. ------- are naturally occurring chemical compound containing basic nitrogen atoms

(A) Alkaloids (B) Tannins
(C) Saponins (D) Phytins

32. -------- are naturally occurring chemical compound containing enough phenolic, hydroxyl and other suitable group

(A) Alkaloids (B) Tannins
(C) Saponins (D) Phytins

33. --------------- are steroid and triterpene glycosides, so named because of their soap like properties

(A) Alkaloids (B) Tannins
(C) Saponins (D) Phytins

34. ---------- are the major form of phosphorous storage in seeds

(A) Alkaloids (B) Tannins
(C) Saponins (D) Phytins

35. In which year ABA was isolated?

(A) 1961 (B) 1962
(C) 1963 (D) 1964

36. Who has isolated ABA in 1963?

(A) Addicott (B) Murashige
(C) Skoog (D) Redenbaugh

37. ---------------- are first extracted from *Fusarium moniliforme*

(A) ABA (B) Gibberellins
(C) Auxins (D) Cytokinins

38. Gibberellins are first extracted from

(A) *Fusarium oxysporum*
(B) *Fusarium solani*
(C) *Fusarium moniliforme*
(D) *Fusarium lingam*

39. An organic substance that can withstand environmental extremes cannot be degraded by any enzyme is

(A) Lignin (B) Cellulose
(C) Cuticle (D) Sporopollenin

40. Translocation of carbohydrates mostly takes place in the form of :

(A) Maltose (B) Sucrose
(C) Fructose (D) Glucose

41. In maize source of high lysine includes:

(A) Opaque 1 (B) Floury 5
(C) Brittle 5 (D) Gammaque 1

42. The toxic substance gossypol is found in:

(A) Castor (B) Cotton
(C) Groundnut (D) Rapeseed

ANSWER SHEET

1. **A**	2. **D**	3. **B**	4. **B**	5. **A**	6. **C**	7. **B**	8. **C**	9. **A**	10. **D**
11. **B**	12. **C**	13. **B**	14. **A**	15. **C**	16. **A**	17. **B**	18. **C**	19. **D**	20. **D**
21. **C**	22. **B**	23. **A**	24. **A**	25. **B**	26. **C**	27. **D**	28. **A**	29. **B**	30. **C**
31. **A**	32. **B**	33. **C**	34. **D**	35. **C**	36. **A**	37. **B**	38. **C**	39. **D**	40. **B**
41. **A**	42. **B**								

Chapter 14

SEED DORMANCY

1. Which one of the following is unfavourable for germination?
   (A) Adequate moisture
   (B) Appropriate temperature
   (C) Suitable light regime
   (D) High temperature
2. Dormancy caused by hard seed coat is overcome by an
   (A) Exposure to abrasion
   (B) Exposure to high temperature
   (C) Exposure to light
   (D) Exposure to cold temperature
3. Which one of the following is a primary dormancy
   (A) Quiescence
   (B) Endogenous dormancy
   (C) Thermodormancy
   (D) Photodormancy
4. Which one of the following is a secondary dormancy
   (A) Quiescence
   (B) Endogenous dormancy
   (C) Exogenous dormancy
   (D) Photodormancy
5. Which one of the following event is a resting stage of seed
   (A) Quiescence
   (B) Endogenous dormancy
   (C) Thermodormancy
   (D) Photodormancy
6. ---------------- dormancy is related to immature embryo
   (A) Quiescence (B) Exogenous
   (C) Endogenous (D) Thermo
7. ----------- dormancy is related to seed coat
   (A) Quiescence (B) Exogenous
   (C) Endogenous (D) Thermo
8. Red light is -------------- nm
   (A) 460-560 (B) 560-660
   (C) 660-760 (D) 760-860
9. Far red light is --------------- nm
   (A) 760-860 (B) 660-760
   (C) 560-660 (D) 460-560
10. For which crop, the dormancy is advantageous?
    (A) Sesamum (B) Niger
    (C) Castor (D) Groundnut
11. Which one of the following seed germinates on the plant itself after reaching maturity
    (A) Jack (B) Guava
    (C) Sapota (D) Pineapple
12. Which one of the following light condition is most suitable for non-dormant nature of seed?
    (A) Red / far – red : high
    (B) Red / far – red : low
    (C) Red / far – red : very low
    (D) Darkness

13. Which one of the following light regime is appropriate for inducing dormancy?

(A) Red / far – red : high
(B) Red / far – red : low
(C) Red / far – red : very low
(D) Darkness

14. Which one of the following part of a seed act as a physical barrier

(A) Cotyledon (B) Embryo
(C) Scutellum (D) Pericarp

15. Which part of a seed contains more inhibitory substance

(A) Cotyledon (B) Embryo
(C) Scutellum (D) Pericarp

16. An example for combined dormancy

(A) Pine (B) Teak
(C) Maple (D) Ficus

17. Which one of the following is a treatment for breaking physical dormancy

(A) Scarification
(B) Stratification
(C) Warm stratification
(D) Cold stratification

18. When pre-moistened seeds are exposed to 20-25°C, the treatment is known as ---------

(A) Scarification
(B) Stratification
(C) Warm stratification
(D) Cold stratification

19. When pre-moistened seeds are subjected to 3-5°C, the treatment is called as

(A) Scarification
(B) Stratification
(C) Abrasion
(D) Warm temperature

20. Which one of the following method is used to break the physiological dormancy?

(A) Scarification
(B) Stratification
(C) Removal of seed coat
(D) Sand scarification

21. The example for cold stratification is

(A) Acacia (B) Albizia
(C) Casuarina (D) Eucalyptus

22. The example for manual scarification is

(A) Eucalyptus (B) Casuarina
(C) Terminalia (D) Pterocarpus

23. Which acid is used for acid scarification?

(A) Dilute KOH
(B) Commerical KOH
(C) Diltue H_2SO_4
(D) Commercial H_2SO_4

24. Clipping off a portion of seed coat is done to break dormancy of which kind of seed?

(A) Large seed (B) Small seed
(C) Minute seed (D) Tiny seed

25. What is the dosage of H_2SO_4 for acid scarification of blackgram seeds?

(A) 50 ml/kg of seed
(B) 100 ml/kg of seed
(C) 150 ml/kg of seed
(D) 200 ml/kg of seed

26. Mention the duration of acid scarification for green gram seeds

(A) 2-3 min. (B) 3-4 min.
(C) 4-5 min. (D) 5-6 min.

27. How long the seeds of *Acacia spp.* is to be scarified in H_2SO_4

(A) 5-10 min. (B) 10-15 min.
(C) 15-20 min. (D) 20-25 min.

28. Which one of the following seeds are bioscarified by sheeps and goats

(A) *Acacia purpea*
(B) *Acacia tomentosa*
(C) *Acacia nilotica*
(D) *Acacia fistula*

29. Which part of the seed is decomposed by microbes to break dormancy

(A) Hard endocarp and exocarp
(B) Hard seed coat and exocarp
(C) Papery seed coat and exocarp
(D) Epicarp and mesocarp

30. Which one of the following seed needs alternate wetting and drying to break seed dormancy

(A) Teak (B) Terminalia
(C) Acacia (D) Albizia

31. What is the optimum concentration of GA used for different crop seeds to break seed dormancy

(A) 100-1000 ppm
(B) 1000-1500 ppm
(C) 1500-2000 ppm
(D) 2000-2500 ppm

32. What is the optimum concentration of KNO_3 recommended by ISTA for breaking seed dormancy

(A) 0.1% (B) 0.2%
(C) 0.3% (D) 0.4%

33. An example for immature embryo type of dormancy is

(A) Fraxinus (B) Terminalia
(C) Tectona (D) Eucalyptus

34. An example for mechanical dormancy is

(A) Fraxinus (B) Terminalia
(C) Tectona (D) Eucalyptus

35. An example for chemical dormancy is

(A) Fraxinus (B) Terminalia
(C) Tectona (D) Eucalyptus

36. An example for photo-dormancy is

(A) Fraxinus (B) Terminalia
(C) Tectona (D) Eucalyptus

37. An example for thermo-dormancy is

(A) Fraxinus (B) Terminalia
(C) Tectona (D) Quercus

38. The condition that is needed by most seeds to break dormancy is

(A) Exposure to heat
(B) Exposure to cold
(C) Abrasion of the seed
(D) Exposure to moisture

39. Seeds in digestive system of animals are

(A) Absorbed in large intestine
(B) Digested later in small intestine
(C) Passed out through anus
(D) Stored in rectum

40. Fruits eaten by animals are

(A) Tasteless and dry
(B) Juicy and bright colored
(C) Juicy but dull colored
(D) Tasteless and dull colored

41. Seed dormancy allows the plants to

(A) Overcome unfavourable climate conditions
(B) Develop healthy seeds
(C) Reduce viability
(D) Prevent deterioration of seeds

42. Among the following which compound can induce seed dormancy?

(A) Abscisic acid (B) Potassium nitrate
(C) Gibberellins (D) Ethylene

43. Seed dormancy may be due to

(A) Immature embryo
(B) Soft seed coat
(C) Absence of germination inhibitors
(D) Mature embryo

44. No matter how favourable environmental conditions, seeds do not germinate it

(A) They have not settled
(B) Neighboring seeds do not germinate
(C) If seeds do not undergo a dormant or resting period
(D) Seeds are soaked in water

45. Fresh ungerminated seeds are usually

(A) Hard seed (B) Dead seed
(C) Dormant seed (D) Decayed seed

46. Which method is followed for testing of xanthomonas?

(A) Dormancy
(B) Quiescence
(C) Primary dormancy
(D) Secondary dormancy

47. The condition in which the seed can germinate upon absorption of water is known as

(A) Dormancy
(B) Quiescence
(C) Primary dormancy
(D) Secondary dormancy

48. Dormancy which develops in the seed where it is attached to the plant and exist after harvest is named as --------------------

(A) Endogenous dormancy
(B) Exogenous dormancy
(C) Primary dormancy
(D) Secondary dormancy

49. When the seeds during germination are exposed to unfavourable environmental conditions, they do not germinate. What will be the mechanisms for this condition?

(A) Endogenous dormancy
(B) Exogenous dormancy
(C) Primary dormancy
(D) Secondary dormancy

50. What may be the reason for non germination of seeds due to immature embryo?

(A) Endogenous dormancy
(B) Exogenous dormancy
(C) Primary dormancy
(D) Secondary dormancy

51. Which one of the following mechanism is responsible for non germination of seed due to presence of inhibitors?

(A) Endogenous dormancy
(B) Exogenous dormancy
(C) Primary dormancy
(D) Secondary dormancy

52. Which one of the following is related the primary dormancy?

(A) Environmental effects
(B) Genotype effects
(C) Environment and genotype effects
(D) Generation effects

53. A classical example for double dormancy is -------------------------.

(A) Acacia (B) Teak
(C) Apple (D) Mango

54. -------------- is an example for mechanical dormancy.

(A) Acacia (B) Teak
(C) Apple (D) Mango

55. An example for double dormancy is --------

(A) Acacia (B) Teak
(C) Apple (D) Mango

56. Mention a crop in which dormancy studies have been conducted extensively.

(A) *Arabidopsis thaliana*
(B) *Oryza sativa*
(C) *Zea mays*
(D) *Sorghum vulgare*

57. Which one of the following is an example for immature embryo?

(A) *Mangifera indica*
(B) *Citrus* spp.
(C) *Annona squamosa*
(D) *Zyzygium* spp.

58. What crop is an example for thermo dormancy?

(A) Lactuca (B) Phacetia
(C) Tectona (D) Acacia

59. What is an example for photo dormancy?

(A) Lactuca (B) Phacetia
(C) Tectona (D) Acacia

ANSWER SHEET

1. **D**	2. **A**	3. **B**	4. **D**	5. **A**	6. **C**	7. **B**	8. **C**	9. **A**	10. **D**
11. **A**	12. **A**	13. **D**	14. **D**	15. **D**	16. **B**	17. **A**	18. **C**	19. **B**	20. **B**
21. **D**	22. **C**	23. **D**	24. **A**	25. **B**	26. **A**	27. **B**	28. **C**	29. **D**	30. **A**
31. **A**	32. **B**	33. **A**	34. **B**	35. **C**	36. **D**	37. **D**	38. **D**	39. **C**	40. **C**
41. **A**	42. **A**	43. **A**	44. **C**	45. **C**	46. **A**	47. **B**	48. **C**	49. **D**	50. **A**
51. **A**	52. **C**	53. **B**	54. **A**	55. **B**	56. **A**	57. **C**	58. **A**	59. **B**	

Chapter 15

SEED GERMINATION

1. ------------- is an example for epigeal germination
(A) Bean (B) Maize
(C) Pearlmillet (D) Rice

2. -------------- is an example for hypogeal germination
(A) Bean (B) Blackgram
(C) Groundnut (D) Rice

3. What is the final product of glycolysis?
(A) Glucose (B) Triose
(C) Pyruvate (D) Malate

4. Which enzyme is used for break down of fat
(A) Amylase (B) Maltase
(C) Dehydrogenase
(D) Lipase

5. What is the end product of lipid metabolism
(A) Glucose (B) Fructose
(C) Galactose (D) Sucrose

6. During seed germination, at what stage the cell division or cell cycle activity occurs
(A) Imbibition
(B) Enzyme activation
(C) Pre-germination
(D) Post germination

7. Which enzyme is used to metabolise carbohydrate
(A) Amylase (B) Lipase
(C) Proteinase (D) Dehydrogenase

8. Mention the enzyme which is involved in protein metabolism.
(A) Amylase (B) Lipase
(C) Proteinase (D) Dehydrogenase

9. Through which part of seed, the water enters into the seed during imbibitions
(A) Hilum (B) Raphe
(C) Funicle (D) Micropyle

10. Which part of seed emerges first during germination?
(A) Plumule (B) Radicle
(C) Cotyledon (D) Scutellum

11. In which part of the seed, the amylase secretes?
(A) Pericarp
(B) Embryonic axis
(C) Aleurone layer
(D) Cotyledon

12. Which factor is not related to the optimum and suitable factors for germination?
(A) Water (B) Oxygen
(C) Temperature (D) Carbon dioxide

13. Which hormone triggers the denovo synthesis of hydrolytic enzyme during imbibitions?
(A) GA (B) ABA
(C) IBA (D) IAA

14. Which one of the following is the hydrolytic enzyme?
(A) Dehydrogenase
(B) Amylase
(C) Peroxidase
(D) Catalase

15. Which composition of seed absorbs more water faster?
(A) Carbohydrate (B) Fat
(C) Protein (D) Vitamin

16. Fleshy adventitious roots are found in
(A) Potato and onion
(B) Radish and carrot
(C) Dahlia and sweet potato
(D) Turnip and rice

17. Plants whose seeds are germinated by following hypogeal germination are
(A) Beans and cotton
(B) Sugarcane and papaya
(C) Coconut and maize
(D) Cotton and jute

18. The root develops from which part of the germinating seed?
(A) Cotyledons (B) Radicle
(C) Pericarp (D) Plumule

19. Cork Cambium is
(A) Pro-meristem
(B) Ground meristem
(C) Lateral meristem
(D) Intercalary meristem

20. Endosperm is consumed by developing embryo in the
(A) Maize (B) Coconut
(C) Castor (D) Pea

21. When cereal grains germinate, the food reserve of the ---------------- are used first.
(A) Scutellum (B) Suspensor
(C) Aleurone layer (D) Endosperm

22. Growth of herbaceous plants comes from the
(A) Apical meristem
(B) Lateral meristem
(C) Vascular cambium
(D) Cork cambium

23. When cereal grains germinate, amylase and other hydrolytic enzymes are secreted by the
(A) Scutellum (B) Suspensor
(C) Aleurone layer (D) Endosperm

24. In angiosperm embryos, the shoot apical meristem may be located at the tips of the
(A) Hypocotyl (B) Plumule
(C) Radicle (D) Epicotyl

25. Which of the following is (are) associated with germination?
(A) Increase in water content
(B) Resumption of catabolism
(C) Intake of CO_2
(D) DNA

26. Plant development includes all of the following except
(A) Seed formation
(B) Tissue formation
(C) Formation of embryo and suspensor
(D) Deteriorating cotyledons during germination

27. Plant development involves the activity of homeotic genes which control
(A) The orientation of the roots and stems
(B) The start of germination
(C) The formation of seeds
(D) The patterned formation of organ tissues

28. Following germination, the emerging plant gains nutrients for metabolism from the surrounding soil
(A) True (B) False

29. The scutellum become the first leaves of the plant.

(A) True (B) False

30. Adventitious roots grow from

(A) Radical
(B) Plumule
(C) Hypophysis
(D) Any part of the plant body other than the radicle

31. Ascending portion of the axis of the plant, developing directly from the plumule and bears leaves, branches and flower is known as

(A) Root
(B) Stem
(C) Modified root
(D) Apical meristem

32. Root bear

(A) Unicellular hairs
(B) Multicellular hairs
(C) Bicellular hairs
(D) None of the above

33. Root has

(A) Root cap
(B) Region of cell division
(C) Give proper storage to plant
(D) Region of light absorption

34. The diffusion of water through a differentially permeable membrane is called

(A) Osmosis
(B) Imbibition
(C) Diffusion
(D) Water movement

35. Vivipary means

(A) Seed germinates inside the fruit while it is still on the parent tree
(B) Seed germinates as usual
(C) Seed dormancy is very long
(D) Because of immature seeds

36. This is the process by which the plant embryo or baby plant within the seed grows and the baby seedling comes forth from the seed.

(A) Pollination
(B) Fertilziation
(C) Germination
(D) Anthesis

37. Types of germination on basis of epicotyl and hypocotyl elongation are

(A) Epigeal germination
(B) Hypogeal germination
(C) Epicotyls germination
(D) Both A and B

38. External conditions on which germination of seeds is dependent on

(A) Sufficient food storage and water
(B) Live embryo
(C) Sufficient food storage
(D) Oxygen and water

39. Plants whose seeds are germinated by following epigeal germination are

(A) Pea and beans
(B) Papaya and cotton
(C) Coconut
(D) Maize

40. Type of germination in which cotyledons remained underground after elongation of epicotyl is called

(A) Epigeal germination
(B) Hypogeal germination
(C) Epicotyls germination
(D) Hypocotyls germination

41. During seed germination, seed coat ruptures due to

(A) Differentiation of cotyledons
(B) Massive glycolysis in endosperm and cotyledons
(C) Sudden increase in cell division
(D) Massive inhibition of water

42. Seed develops from

(A) Ovary (B) Embryo
(C) Ovule (D) Embryo sac

43. An albuminous seed showing hypogeal germination is

(A) Castor (B) Bean
(C) Gram (D) Maize

44. Proteinaceous part of maize endosperm is

(A) Apophysis
(B) Scutellum
(C) Aleurone layer
(D) Peripheral layer

45. Vivipary is

(A) Seed germination with subterranean cotyledons
(B) Seed germination with epiterranean cotyledons
(C) Fruit development without pollination
(D) Seed germination inside the fruit the fruit while attached to the plant

46. A gas required for germination of pea seed is

(A) Nitrogen (B) Oxygen
(C) Hydrogen (D) Water vapours

47. Protective covering over radical during seed germination is

(A) Suspensor (B) Coleorhiza
(C) Epithelium (D) Coleoptile

48. Germination is epigeal in

(A) Rice (B) Wheat
(C) Maize (D) Sunflower

49. First step in seed germination is

(A) Passing through gut of an animal
(B) Exploding along sutures
(C) Absorption of water
(D) Taking in of O_2

50. With an increase in seed size upon germination, dry mass of seeds

(A) Increases
(B) Decreases
(C) Vanishes
(D) Grows proportionally with seed

51. First part to come out of seed is the

(A) Radicle (B) Plumule
(C) Stem (D) Cotyledon

ANSWER SHEET

1. **A**	2. **D**	3. **C**	4. **D**	5. **D**	6. **D**	7. **A**	8. **C**	9. **D**	10. **B**
11. **C**	12. **D**	13. **A**	14. **B**	15. **C**	16. **B**	17. **C**	18. **B**	19. **C**	20. **D**
21. **A**	22. **A**	23. **C**	24. **D**	25. **A**	26. **D**	27. **D**	28. **A**	29. **A**	30. **D**
31. **D**	32. **A**	33. **A**	34. **A**	35. **A**	36. **C**	37. **D**	38. **D**	39. **B**	40. **B**
41. **D**	42. **C**	43. **D**	44. **C**	45. **D**	46. **B**	47. **D**	48. **D**	49. **C**	50. **A**
51. **A**									

Chapter 16

HORMONES

1. Coconut milk factor is
   (A) An auxin (B) A gibberelin
   (C) Abscisic acid (D) Cytokinin
2. Which one of the followings is a synthetic auxin?
   (A) IAA (B) NAA
   (C) IBA (D) GA
3. An enzyme that can stimulate germination of barley seeds is
   (A) α-amylase (B) Lipase
   (C) Protease (D) Invertase
4. Seed dormancy is due to the:
   (A) Ethylene (B) Abscisic acid
   (C) IAA (D) Starch
5. Importance of day length in flowering of plants was first shown in:
   (A) Petunia (B) Letting
   (C) Tobacco (D) Cotton
6. Photographic curvature is the result of uneven distribution of:
   (A) Auxin (B) Gibberellin
   (C) Phytochrome (D) Cytokinins
7. Natural cytokinins are synthesized in tissues that are:
   (A) Senescent
   (B) Dividing rapidly
   (C) Storing food
   (D) Differentiating material
8. Which one of the followings is a gaseous plant hormone?
   (A) Ethylene
   (B) Gibberellin
   (C) IAA
   (D) Abscisic acid
9. Differentiation of shoot is controlled by:
   (A) High auxin : cytokinin ratio
   (B) High cytokinin : auxin ratio
   (C) High gibberellins : auxin ratio
   (D) High gibberellin : cytokinin ratio
10. Opening of buds into flowers, is a type of:
    (A) Autonomic movement of variation
    (B) Autonomic movement of locomotion
    (C) Autonomic movement of growth
    (D) Paratonic movement of growth
11. Which one of the following acids is a derivative of carotenoids?
    (A) Abscisic acid
    (B) Indole butyric acid
    (C) Indole-3-acetic acid
    (D) Gibberellic acid
12. Cell elongation in internodal regions of the green plants takes place due to:
    (A) Cytokinins
    (B) Gibberellins
    (C) Ethylene
    (D) Indole acetic acid

13. Which one of the following pairs, is not correctly matched?

(A) IAA – Cell elongation

(B) Abscisic acid – Stomatal closure

(C) Gibberellic acid – Leaf fall

(D) Cytokinin – Cell division

GIBBERELLIN

14. Gibberellins was named after a fungus called

(A) *Gibberella fujikuroi*

(B) *Gibberella gaditijirrii*

(C) *Gibberella acuminate*

(D) *Gibberella africana*

15. Gibberellins are chemically

(A) Phenolic derivatives

(B) Terpenoid derivatives

(C) Adenine derivatives

(D) Alkaloid derivatives

16. *Gibberella fujikuroi* the fungus causes

(A) Foolish seedling disease of rice or bakanae disease

(B) Damping off seedling disease of rice or bakanae disease

(C) Fungal blight disease of rice or bakanae disease

(D) Rust disease of rice or bakanae disease

17. The most studied gibberellins is

(A) GA_1 (B) GA_9

(C) GA_3 (D) GA_4

18. All are anti gibberellins except

(A) Acetosyringone

(B) Amo1618

(C) Maleic hydrazide

(D) Phosphon D

19. All are functions of gibberellins except

(A) Promote cell elongation

(B) Promote dormancy

(C) Promote elongation dormancy

(D) Promotes parthenocarpic fruit formation

20. Which of the following bioassay is used to test the presence of gibberellins

(A) Chlorophyll preservation test

(B) Dwarf pea elongation technique

(C) Rice seedling growth inhibition test

(D) Excised radish cotyledon enlargement test

21. The gibberellins present in the aleurone layer of barley increase transcription of

(A) Lipase and protease

(B) Protease and cellulose

(C) Protease and amylase

(D) Lipase and cellulose

22. Elongation of rosette plants is achieved by treatment with

(A) Auxin (B) Zeatin

(C) Cytokinin (D) Gibberellin

23. Translocation of gibberellin is non-polar and predominantly takes place through

(A) Xylem

(B) Phloem

(C) Both phloem and xylem

(D) Capillary rise

24. The hormone which is involved in regulation of florigen synthesis

(A) Ethylene (B) Zeatin

(C) Cytokinin (D) Gibberellin

25. Which is the gibberellin involved in elongation responses

(A) GA_1 (B) GA_9

(C) GA_3 (D) GA_4

26. Which is the gibberellin involved in flowering

(A) GA_1 (B) GA_9

(C) GA_3 (D) GA_4

27. Early stages of gibberellins synthesis take place in

(A) Mitochondria (B) Peroxisome
(C) Plastids (D) Cytoplasm

28. Genetic dwarfism can be nullified by spraying with

(A) Auxin (B) Zeatin
(C) Cytokinin (D) Gibberellin

AUXINS

29. Auxins

(A) Promote apical dominance
(B) Promote fruit growth
(C) Induce parthenocarpy
(D) All of above

30. Auxins promote growth at a very low concentration in

(A) Shoots (B) Roots
(C) Flowers (D) Nodes

31. Auxins tend to inhibit

(A) Fruit growth
(B) Parthenocarpy
(C) Abscission
(D) Bud initiation

32. In apical, dominance action of auxins is enhanced by

(A) Ethane (B) Oxytocin
(C) Gibberellin (D) Cytokinins

33. The term auxin was coined by

(A) Skoog (B) Haberlandt
(C) Miescher (D) F.W.Went

34. Precursor of indole acetic acid (natural auxin) is

(A) Glycine
(B) Methionine
(C) Isopentynyl Pyrophosphate
(D) Tryptophan

35. All the following hormones are growth inhibitors except

(A) Abscisic (B) Dorman
(C) Ethylene (D) IAA

36. Which of the following is a gaseous hormone

(A) IBA (B) NAA
(C) Abscisic acid (D) Ethylene

37. Which of the following is widely used as a rooting hormone

(A) NAA
(B) 2, 4-D
(C) 2, 4, 5-T
(D) Cytokinin

38. "Agent Orange" the leaf defoliator used by USA in Vietnam war was

(A) Ethylene
(B) 2,4-D and 2,4,5-T
(C) 2,4-D and NAA
(D) 2,4,5-T, NAA and ethylene

39. All the following inhibits auxin transport (anti-auxins) except

(A) Cytokinin
(B) α-naphthyl thalamic acid
(C) 2,3,5 – tri iodo benzoic acid
(D) Ethylene

40. Which of the following auxin is widely used as a selective weedicide

(A) IBA (B) 2,4-D
(C) NAA (D) 2,4,5-T

41. All are synthetic auxins except

(A) NAA (B) IAA
(C) 2,4-D (D) 2,4,5-T

42. Auxin transport is

(A) Polar
(B) Non-polar
(C) Symplast
(D) Apoplast

43. Which of the following bioassays are used to detect the presence of auxin

(A) Avena curvature test and tobacco pith culture

(B) Split pea stem curvature test and tobacco pith culture

(C) Avena curvature test and split pea stem curvature test

(D) Tobacco pith culture only

44. Which of the following is not a function of auxin

(A) Inducing dormancy

(B) Enhancing cell division

(C) Inducing callus formation

(D) Maintaining apical dominance

45. Nodule formation is induced by

(A) IBA (B) IAA

(C) Both (A) and (B)

(D) NAA

46. All are natural auxins except

(A) IAA (B) Phenoxy acetic acid

(C) Indole-3-acetic acid

(D) NAA

47. Which of the following is an auxin receptor

(A) ETR1 (B) CBP1

(C) ABP1 (D) GRE

CYTOKININS

48. Cytokinins are

(A) Adenine derivatives

(B) Guanine derivatives

(C) Cytidine derivatives

(D) Thymine derivatives

49. Which of the following is a natural cytokinin

(A) Isopentanyl adenine

(B) Zeatin

(C) 6-isopentanyl adenine

(D) 6-amino purine

50. The first kintin isolated by miller was from

(A) Coconut endosperm

(B) Maize grains

(C) Herring sperm DNA

(D) Wheat grains

51. The first natural cytokinin isolated from liquid endosperm was

(A) Isopentanyl adenine

(B) Zeatin

(C) 6-isopentanyl adenine

(D) 6-amino purine

52. *Agrobacterium tumefaciens* in plasmids t-DNA has genes for

(A) Auxin and gibberellins

(B) Cytokinins and ethylene

(C) Cytokinins and gibberellins

(D) Auxin and cytokinins

53. Which of the following is the precursor of cytokinins

(A) Adenine (B) Guanine

(C) Cytidine (D) Thymine

54. Cytokinins are predominantly present in

(A) Permanent tissues

(B) Meristamatic tissues

(C) Endodermis

(D) Cortical region

55. All are bioassays for cytokinins except

(A) Chlorophyll preservation test

(B) Tobacco pith culture

(C) Rice seedling growth inhibition test

(D) Excised radish cotyledon enlargement test

56. All statement are true regarding cytokinin except

(A) Promote cell division

(B) Delay senescence

(C) Induce dormancy

(D) Counteract apical dominance

57. Cytokinin treatment enhances
(A) Chlorophyll synthesis
(B) Chlorophyll degradation
(C) Protein degradation
(D) Water uptake

58. In tissue culture, high cytokinins to auxin ratio causes
(A) Root differentiation
(B) Shoot differentiation
(C) Both A and B
(D) None of these

59. In tissue culture low cytokinins to auxin ratio causes
(A) Root differentiation
(B) Shoot differentiation
(C) Both A and B
(D) None of these

60. Translocations of cytokinin is polar and takes place through
(A) Xylem (B) Phloem
(C) Both phloem and xylem
(D) Capillary rise

61. "Pomalin" is effective in increasing apple size and is a combination of
(A) Auxin and gibberellins
(B) Cytokinins and ethylene
(C) Cytokinins and ethylene
(D) Auxin and cytokinins

62. Which of the following hormone is involved in counteracting apical dominance induced by auxin
(A) Cytokinin (B) Ethylene
(C) Abscisic acid (D) Brassinosteroids

ETHYLENE

63. Which of the following is a naturally occurring gaseous hormone
(A) IAA (B) Zeatin
(C) ZiP (D) Ethylene

64. High concentration of ethylene is present in
(A) Young leaves
(B) Meristamatic regions
(C) Buds (D) Ripening fruits

65 The precursor of eth r33333cccccccccc2Rylene is
(A) Tryptophan (B) Serine
(C) Methionine (D) Cysteine

66. All the following are functions of ethylene except
(A) Promote senescence
(B) Enhances abscission
(C) Promote ripening
(D) Induces cell division

67. All are ethylene inhibitors except
(A) Phenyl mercuric acetate
(B) Ag^+
(C) Aminooxy acetic acid
(D) Aminoethoxy vinyl glycine

68. Which of the following ion is an ethylene inhibitor
(A) $Mg2^+$ (B) NH_3^+
(C) Ag^+ (D) Cl^-

69. Which of the following is a bioassay for ethylene
(A) Chlorophyll preservation test
(B) Triple pea test
(C) Rice seedling growth inhibition test
(D) Excised radish cotyledon enlargement test

70. Climacteric fruits are
(A) Fruits exhibiting increase in respiration before final stages of ripening
(B) Fruits exhibiting decrease in respiration before final stages of ripening
(C) Fruits exhibiting increase in photosynthesis before final stages of ripening
(D) Fruits exhibiting decrease in photosynthesis before final stages of ripening

71. All are climacteric fruits that respond to ethylene except

(A) Apple (B) Banana
(C) Grape (D) Mango

72. Aerenchyma formation in wet land species especially in rice is induced by

(A) Auxin and ethylene
(B) Cytokinin
(C) ABA (D) Ethylene

73. Which of the following enzyme production is enhanced by ethylene

(A) Amylase and protease
(B) Protease only
(C) Cellulase and chlorophyllase
(D) Cellulase and protease

74. Which of the following is an artificial ripening agent

(A) Ethaphone (B) 2,4-D
(C) NAA (D) Ethylene

75. Which of the following hormone stimulates ethylene release

(A) Auxin (B) Cytokinin
(C) Gibberellin (D) Abscisic acid

76. Ethylene production by organisms can be detected by

(A) Coloum chromatography
(B) Gas chromatography
(C) Spectrophotometer
(D) Colorimeter

POLYAMINES

77. Polyamines are compounds containing

(A) One amine group
(B) Two or more amine group
(C) One amide group
(D) Two or more amide group

78. Precursors of polyamines are

(A) Arginine and asparagines
(B) Arginine and glutamine
(C) Arginine and lysine
(D) Arginine and serine

79. All the following are polyamines except

(A) Spermidine
(B) Spermine
(C) Putrescine
(D) Lysergine

80. Brassinosteroids are lipids synthesized from

(A) Cholesterol (B) Stigmasterol
(C) Campesterol (D) Ergosterol

81. Brassinosteroids are involved in enhancing

(A) Senescence (B) Dormancy
(C) Cell division (D) Ripening

82. Which of the following is a synthetic growth inhibitor

(A) Ethylene (B) ABA
(C) Morphactin (D) IBA

83. Which of the following is a flowering hormone

(A) Ethylene (B) ABA
(C) Morphactin (D) Florigen

84. The hormone that is produced during chilling treatment?

(A) IAA (B) Ethylene
(C) Gibberellin (D) Vernalin

85. Traumatic acid is a

(A) Stress hormone
(B) Flowering hormone
(C) Wound hormone
(D) Growth hormone

86. Which of the following is a hormone derived from phenols

(A) Morphactins
(B) Phosphon D
(C) Polyamines
(D) Coumarin

87. Which of the following are flowering hormones
(A) Vernalin and florigen
(B) Florigen and ethylene
(C) Ethylene only
(D) both a and b

88. Caulines are hormones
(A) Specific to particular region of a plant
(B) Of meristamatic region
(C) Of abscission layers
(D) None of these

ABSCISIC ACID

89. Which of the following hormone is a growth inhibitor
(A) IAA (B) 2ip
(C) GA_3 (D) Dormin

90. The active form of ABA is
(A) (+) cis ABA
(B) (+) 2-trans ABA
(C) (+) cis 2 trans ABA
(D) (+) trans ABA

91. The precursor of ABA is
(A) Violaxanthin (B) Methionine
(C) Tryptophan (D) Carotenoids

92. Which of the following is a bioassay for ABA
(A) Chlorophyll preservation test
(B) Triple pea test
(C) Rice seedling growth inhibition test
(D) Excised radish cotyledon enlargement test

93. All the following are functions of ABA except
(A) Induce dormancy
(B) Enhances abscission
(C) Involved in stress response
(D) Cell elongation

94. ABA stimulates release of
(A) Auxin (B) Ethylene
(C) Cytokinin (D) GA

95. During water stress rise in ABA levels initially cause
(A) Stomatal opening
(B) Stomatal closure
(C) Reduced transpiration
(D) Increased transpiration

96. Transport of ABA takes place through
(A) Xylem (B) Phloem
(C) Endodermis (D) Both A and B

97. ABA is an
(A) Anti-Auxin
(B) Anti-Cytokinin
(C) Anti-Gibberellin
(D) Ethylene inhibitor

98. Which of the following hormone functions as an anti-transpirant
(A) Auxin (B) Cytokinin
(C) ABA (D) Ethylene

99. ABA is also called as
(A) Stress hormone
(B) Ripening hormone
(C) Growth hormone
(D) None of these

100. High levels of ABA are synthesized in
(A) Tissues undergoing cell division
(B) Tissues undergoing cell elongation
(C) Tissues undergoing stress
(D) Tissues undergoing ripening

101. ABA synthesis begins in
(A) Cytoplasm and chloroplast
(B) Chloroplast and amyloplast
(C) Cytoplasm and amyloplast
(D) Chloroplast and amyloplast

102. Gibberellins are now commercially used into increase sugarcane growth and sugar yields :

(A) Japan
(B) India
(C) Hawaii
(D) Korea

103. There is an increase in ------------ under drought conditions.

(A) ABA content
(B) Ethylene level
(C) Proline content
(D) All of these

ANSWER SHEET

1. **D**	2. **B**	3. **A**	4. **B**	5. **C**	6. **A**	7. **B**	8. **A**	9. **B**	10. **C**
11. **A**	12. **B**	13. **C**	14. **A**	15. **B**	16. **A**	17. **C**	18. **A**	19. **B**	20. **B**
21. **C**	22. **D**	23. **B**	24. **D**	25. **A**	26. **B**	27. **C**	28. **D**	29. **D**	30. **B**
31. **C**	32. **C**	33. **D**	34. **D**	35. **D**	36. **D**	37. **A**	38. **B**	39. **A**	40. **B**
41. **B**	42. **A**	43. **C**	44. **A**	45. **B**	46. **D**	47. **C**	48. **A**	49. **B**	50. **C**
51. **B**	52. **D**	53. **A**	54. **B**	55. **C**	56. **C**	57. **A**	58. **B**	59. **A**	60. **A**
61. **C**	62. **A**	63. **D**	64. **D**	65. **C**	66. **D**	67. **A**	68. **C**	69. **B**	70. **A**
71. **C**	72. **D**	73. **C**	74. **A**	75. **D**	76. **B**	77. **B**	78. **C**	79. **D**	80. **C**
81. **C**	82. **C**	83. **D**	84. **D**	85. **C**	86. **D**	87. **A**	88. **A**	89. **D**	90. **A**
91. **A**	92. **C**	93. **D**	94. **B**	95. **B**	96. **D**	97. **C**	98. **C**	99. **C**	100. **C**
101. **D**	102. **C**	103. **D**							

Unit 2
SEED PRODUCTION

Chapter 1

HETEROSIS AND HYBRID VIGOUR

1. How to estimate the relative heterosis?

   (A) $[(F_1 - AP) / AP] \times 100$
   (B) $[(F_1 - BP) / BP] \times 100$
   (C) $[(F_1 - MP) / MP] \times 100$
   (D) $[(F_1 - SH) / SH] \times 100$

2. ------------ formula is used for estimation of Heterobeltiosis.

   (A) $[(F_1 - AP) / AP] \times 100$
   (B) $[(F_1 - BP) / BP] \times 100$
   (C) $[(F_1 - MP) / MP] \times 100$
   (D) $[(F_1 - SH) / SH] \times 100$

3. Which from the following is the formula for estimation of standard heterosis?

   (A) $[(F_1 - AP) / AP] \times 100$
   (B) $[(F_1 - BP) / BP] \times 100$
   (C) $[(F_1 - MP) / MP] \times 100$
   (D) $[(F_1 - SH) / SH] \times 100$

4. Which one of the following is a maintainer line?

   (A) A line (B) B line
   (C) R line (D) M line

5. Which one of the following is an isogenic line?

   (A) A line (B) B line
   (C) R line (D) M line

6. Which one of the following is a male sterile line?

   (A) A line (B) B line
   (C) R line (D) M line

7. Which from the following is a male line of a three line hybrid seed production?

   (A) A line (B) B line
   (C) R line (D) M line

8. Which one of the following is a single cross hybrid?

   (A) $A \times R = F_1$
   (B) $(A \times B) \times (C \times D) = F_1$
   (C) $(A \times B) \times C = F_1$
   (D) $(A \times B) \times OPV = F_1$

9. Which one of the following is a double cross hybrid?

   (A) $A \times R = F_1$
   (B) $(A \times B) \times (C \times D) = F_1$
   (C) $(A \times B) \times C = F_1$
   (D) $(A \times B) \times OPV = F_1$

10. Which one of the following is a three way cross hybrid?

    (A) A ----x $R = F_1$
    (B) $(A \times B) \times (C \times D) = F_1$
    (C) $(A \times B) \times C = F_1$
    (D) $(A \times B) \times OPV = F_1$

11. Which one of the following is a top cross hybrid?

    (A) $A \times R = F_1$
    (B) $(A \times B) \times (C \times D) = F_1$
    (C) $(A \times B) \times C = F_1$
    (D) $(A \times B) \times OPV = F_1$

12. Heterosis is
(A) Appearance of spontaneous mutations
(B) Induction of mutations
(C) Mixture of two or more traits
(D) Superiority of hybrids over their parents

13. The new varieties of plants are produced by
(A) Introduction and mutation
(B) Selection and hybridisation
(C) Mutation and selection
(D) Selection and introduction

14. Pure line breed refers to
(A) Heterozygosity only
(B) Homozygosity only
(C) Homozygosity and self assortment
(D) Heterozygosity and linkage

15. Heterosis over the better parent is called as
(A) Relative heterosis
(B) Standard heterosis
(C) Pseudoheterosis
(D) Heterobeltiosis

16. How many cells will be produced from a cell if the cell divides meiotically?
(A) 1 (B) 2
(C) 3 (D) 4

17. If a character is cytoplasmically controlled then
(A) The progeny resembles female parent
(B) The progeny resembles male parent
(C) It's segregation of 3:1 is observed in F1
(D) 15:1 segregation in F1

18. If a dihybrid is test crossed, the phenotypic ratio of progeny will be
(A) 9:3:3:1 (B) 9:3:4
(C) 1:1:1:1 (D) 15:1

19. If a dihybrid without genic interaction is selfed the following ratio is obtained in F2
(A) 1:1:1:1 (B) 9:3:3:1
(C) 9:7 (D) 9:6:1

20. In which one of the following crops, maximum exploitation of hybrid vigour during the last three decades has been done?
(A) Wheat (B) Rice
(C) Maize (D) Cotton

21. Multiline varieties are mixtures of
(A) Several morphologically different plants
(B) Bulked progenies of different inbred having identical disease resistance genes lines
(C) Several pure lines of similar phenotype
(D) Morphologically identical plants with genes for disease resistance different genetic make-up

22. Progeny of a hybrid plant
(A) Breds true
(B) Resembles maternal plant
(C) Resembles paternal plant
(D) Segregation

23. Selfing of a plant increases
(A) Heterozygosity
(B) Homozygosity
(C) Homogenecity
(D) Heterogenecity

24. The first generation progeny of a hybrid are called
(A) Pure lines (B) Segregates
(C) Parents (D) F1 plants

25. The production of double cross hybrid in maize was first suggested by
(A) Shull (B) East
(C) Jones (D) Hull

26. The term heterosis was first used by
(A) Mendel (B) Shull
(C) Bateson (D) Punnet

27. Theory of overdominance states

(A) Dominant homozygote is superior to recessive homozygote

(B) Recessive homozygote is superior to dominant homozygote

(C) Heterozygote is superior to both the homozygote

(D) Elimination of unfavourable variations

28. When an F1 is crossed with the recessive homozygote parent the cross is known as

(A) Back cross (B) Top cross

(C) Test cross (D) Single cross

29. When an individual is having both the alleles of contrasting characters, it is said to be

(A) Monoecious (B) Homozygous

(C) Dioecious (D) Heterozygous

ANSWER SHEET

1. **B**	2. **C**	3. **D**	4. **B**	5. **B**	6. **A**	7. **C**	8. **A**	9. **B**	10. **C**
11. **D**	12. **D**	13. **B**	14. **B**	15. **D**	16. **D**	17. **A**	18. **C**	19. **B**	20. **C**
21. **C**	22. **D**	23. **B**	24. **B**	25. **C**	26. **B**	27. **C**	28. **C**	29. **D**	

Chapter 2

DEVELOPMENT OF HYBRIDS AND RELEASE OF VARIETIES

1. Which one of the following maize variety is a composite variety?

(A) African tall (B) Deccan

(C) Ganga (D) COH1

2. What is the percentage of sterility in GMS?

(A) 100% (B) 50:50%

(C) 25:75% (D) 75:25%

3. Where MLT is conducted?

(A) KVK's of SAU's

(B) Research stations of SAU's

(C) Central Agricultural Universities

(D) ICAR Institutes

4. In variety release, the best entries are proposed for which trial?

(A) RYT (B) PYT

(C) MLT (D) ART

5. For variety release, DUS test is must. Which authority is conducting DUS test?

(A) ICAR (B) ICRISAT

(C) PPV & FRA (D) IARI

ANSWER SHEET

1. **A** 2. **B** 3. **B** 4. **D** 5. **C**

Chapter 3

DETERIORATION OF VARIETIES

1. Who has listed out the factors of apparent and real deterioration of varieties

   (A) Kadam (1942)
   (B) Bateman (1947)
   (C) Dettann (1953)
   (D) Martin (1942)

2. Who has suggested the factors that are responsible for genetic contamination in natural fields due to natural crossing

   (A) Kadam (1942)
   (B) Bateman (1947)
   (C) Dettann (1953)
   (D) Martin (1942)

3. Which one of the following is not relevant for deterioration of varieties due to natural crossing

   (A) Natural crossing with desirable plants
   (B) Natural crossing with undesirable plants
   (C) Natural crossing with diseased plants
   (D) Natural crossing with off-type plants

4. In which crop, the natural crossing is the major source of genetic contamination

   (A) Self fertilized crops
   (B) Cross fertilized crops
   (C) Self compatible crops
   (D) Self incompatible crops

ANSWER SHEET

1. **A** 2. **B** 3. **A** 4. **B**

Chapter 4

SEED GENERATION SYSTEM

1. Which one of the following is four recognised classes of seed?
   (A) Nucleus seed, breeder seed, foundation seed and certified seed
   (B) Nucleus seed, breeder seed, foundation seed – I and certified seed
   (C) Nucleus seed, breeder seed, foundation seed and certified seed – I
   (D) Breeder seed, foundation seed, registered seed and certified seed

2. Golden yellow tag is used for ---------------- seed
   (A) Nucleus (B) Breeder
   (C) Foundation (D) Certified

3. White tag is used for ---------------- seed
   (A) Nucleus (B) Breeder
   (C) Foundation (D) Certified

4. Azhur blue tag is used for ------------------ seed
   (A) Nucleus (B) Breeder
   (C) Foundation (D) Certified

5. Which from the following is the progeny of nucleus seed?
   (A) Certified seed
   (B) Foundation seed
   (C) Registered seed
   (D) Breeder seed

6. Which category of the following is the progeny of foundation seed?
   (A) Certified seed
   (B) Foundation seed
   (C) Registered seed
   (D) Breeder seed

7. Which category of the following is the progeny of breeder seed?
   (A) Certified seed (B) Foundation seed
   (C) Registered seed
   (D) Breeder seed

8. Genetic purity for nucleus seed is --------%
   (A) 100 (B) 99.9
   (C) 99.5 (D) 99

9. Genetic purity for breeder seed is ---------%
   (A) 100 (B) 99.9
   (C) 99.5 (D) 99

10. Genetic purity for foundation seed is -----%
    (A) 100 (B) 99.9
    (C) 99.5 (D) 99

11. Genetic purity for certified seed is -------%
    (A) 100 (B) 99.9
    (C) 99.5 (D) 99

12. Opal green colour tag is used for
    (A) Breeder seed
    (B) Truthful labelled seed
    (C) Foundation seed
    (D) Certified seed

13. Nucleus seed is
    (A) 100% pure (B) 80% pure
    (C) 50% pure (D) 40% pure

14. Registered seed is directly produced from
(A) Foundation seed
(B) Certified seed
(C) Breeder's seed
(D) All the above

15. Breeder seed is the progeny of :
(A) Foundation seed (B) Registered seed
(C) Nucleus seed (D) Certified seed

16. Certification is not required for :
(A) Truthful labelled seed
(B) Breeder seed
(C) Foundation seed
(D) Certified seed

17. Improved seed includes
(A) Nucleus seed
(B) Breeder seed
(C) Foundation seed
(D) All of the above

18. Initial seed of an improved variety is called
(A) Nucleus seed (B) Breeder seed
(C) Foundation seed (D) Certified seed

19. Progeny of a nucleus seed is referred as
(A) Certified seed (B) Foundation seed
(C) Registered seed (D) Breeder seed

20. Seed certification requires
(A) An improved variety
(B) Genetic purity
(C) Physical purity
(D) All of the above

21. Seed meant for generation distribution to the farmers for commercial crop production refers to
(A) Foundation seed
(B) Breeder seed
(C) Certified seed
(D) Nucleus seed

22. International Crop Improvement Association (ICIA) in ----------------- classified seed into different categories
(A) 1964 (B) 1946
(C) 1963 (D) 1972

23. Which of the following is not a commercial class of seed?
(A) Producer (B) Registered
(C) Breeder (D) Foundation

24. Which of the commercial classes of seed is sold to farmers
(A) Registered (B) Breeder
(C) Certified (D) Foundation

25. Nucleus seed is :
(A) 100% pure (B) 30% pure
(C) 50% pure (D) 60% pure

26. Improved seed distributed to farmers for commercial cultivation is :
(A) Nucleus (B) Breeder
(C) Foundation (D) Certified

27. Seed multiplication ratio in paddy is :
(A) 1 : 10 (B) 1 : 40
(C) 1 : 80 (D) 1 : 100

28. Genetically most pure seed is
(A) Truthful labelled
(B) Breeder seed
(C) Certified seed
(D) Foundation seed

ANSWER SHEET

1. **A**	2. **B**	3. **C**	4. **D**	5. **D**	6. **A&C**	7. **B**	8. **A**	9. **B**	10. **C**
11. **D**	12. **B**	13. **A**	14. **A**	15. **C**	16. **B**	17. **D**	18. **A**	19. **D**	20. **D**
21. **C**	22. **B**	23. **A**	24. **C**	25. **A**	26. **D**	27. **C**	28. **B**		

Chapter 5

SEED QUALITY

1. Which one of the following is not relevant with respect to seed quality?
   (A) Embryonic plant
   (B) Non-living organism
   (C) A dispersal unit
   (D) Beginning and end of plant life

2. Which one of the following is not relevant with respect to seed quality?
   (A) Carrier of new technology
   (B) Basic tool for food production
   (C) Principal means for higher yield
   (D) Farm saved seed

3. Which one of the following point is not a physiological quality of seed?
   (A) Shape of seed
   (B) Germination of seed
   (C) Germination rate
   (D) Seed vigour

4. What is the safe moisture for cereal seed?
   (A) 15% (B) 14%
   (C) 12% (D) 18%

5. Which one of the following is a designated inseparable other crop seed for wheat?
   (A) Oats (B) Pearlmillet
   (C) Rye (D) Sorghum

6. Which one of the following is a designated inseparable other crop seed for barley?
   (A) Wheat (B) Pearlmillet
   (C) Rye (D) Sorghum

7. What is the yield increase if you use certified seed for sowing
   (A) 5-10% (B) 10-15%
   (C) 15-20% (D) 20-25%

8. In wheat and rice the genetic purity of a variety can deteriorate due to :
   (A) Inbreeding
   (B) Mechanical contamination with seeds of other varieties
   (C) Delayed segregation
   (D) Due to (A) and (B) both

ANSWER SHEET

1. B 2. D 3. A 4. C 5. A 6. A 7. C 8. A

Chapter 6

SEED POLICY

1. When new seed policy came into force in India?

   (A) 1984 (B) 1986
   (C) 1988 (D) 1982

2. When the National Seed Policy was implemented?

   (A) 2000 (B) 2001
   (C) 2002 (D) 2003

3. When breeder seed policy was introduced?

   (A) 1982 (B) 1986
   (C) 1984 (D) 1988.

4. Which one of the following country, Taikii seed company belongs to?

   (A) Denmark (B) Switzerland
   (C) Germany (D) Japan

5. In which one of the following country, DLF-Trifolium is located?

   (A) Denmark (B) Switzerland
   (C) Germany (D) Japan

6. In which one of the following country, Sakata seed company is situated?

   (A) Denmark (B) Switzerland
   (C) Germany (D) Japan

7. Which one of the following is not applicable for seed quality control system?

   (A) Seed Certification Agency
   (B) Seed Law Enforcement Wing
   (C) Seed Testing Laboratory
   (D) Soil Testing Laboratory

ANSWER SHEET

1. **C** 2. **C** 3. **B** 4. **D** 5. **A** 6. **D** 7. **D**

Chapter 7

SEED PLANNING

1. How SMR is calculated?

   (A) Ratio of seed yield and seed rate
   (B) Ratio of seed rate and seed yield
   (C) Ratio of grain yield and seed rate
   (D) Ratio of seed rate and grain yield

2. How SRR is worked out?

   (A) (Seed yield / Seed rate) x 100
   (B) Seed yield / Seed rate
   (C) (Qty of actual quality seed sown / Qty of quality seed required) x 100
   (D) (Qty of certified seed required / Qty of certified seed sown) x 100

3. What is the SMR for wheat?

   (A) 1:49 (B) 1:150
   (C) 1:120 (D) 1:200

4. What is the SMR for sesamum?

   (A) 1:49 (B) 1:150
   (C) 1:120 (D) 1:200

5. What is the SMR for jute?

   (A) 1:49 (B) 1:150
   (C) 1:120 (D) 1:200

6. What is the SMR for pigeon pea

   (A) 1:49 (B) 1:150
   (C) 1:120 (D) 1:200

7. What is the seed replacement rate for hybrids?

   (A) 25% (B) 50%
   (C) 75% (D) 100%

8. ----------------- % is replacement rate in India

   (A) 10-15 (B) 15-20
   (C) 20-25 (D) 25-30

9. Which one of the following is not applicable for calculating the seed production requirement while seed planning?

   (A) Seed replacement rate
   (B) Planting ratio
   (C) Seed rate of A, B and R lines
   (D) Seed yield of A, B and R lines

10. A farmer wants to cover an area of 2000 ha under certified seed production. The seed rate is 20 kg/ha and seed yield is 1000 kg/ha. What will be quantity of certified seed needed to cover the above area

    (A) 20,000 kg (B) 30,000 kg
    (C) 40,000 kg (D) 50,000 kg

ANSWER SHEET

1. A 2. C 3. A 4. D 5. C 6. B 7. D 8. B 9. A 10. C

Chapter 8

SEED VILLAGE

1. Who has coined the conept of seed village as 'Win-Win Model'?

   (A) Deepak Sharma
   (B) Deepak Verma
   (C) Deepak Sumit
   (D) Deepak Mullick

2. Which organisation is a leading promotor of seed village approach in India?

   (A) CIKS (B) IARI
   (C) MSSRF (D) ICRISAT

3. Where MSSRF is located?

   (A) Chennai (B) Hyderabad
   (C) Bangalore (D) Mumbai

4. In which state, Ranebennur, well known place for hybrid and flower seed production is located?

   [illegible] nil Nadu
   [illegible] nataka
   [illegible] Pradesh
   [illegible] ungana

5. In Tamil Nadu, MSSRF has implemented seed village very successfully. Where it was implemented?

   (A) Kannivadi of Dindigul district
   (B) Andipatti of Theni district
   (C) Udumalpet of Tirupur district
   (D) Melur of Madurai district

6. Which one of the following is not relevant for seed village with respect to its advantages?

   (A) No isolation problem
   (B) Easy mechanization
   (C) Engage trained labourers
   (D) Not easy for inspection by govt. agencies

7. Which one of the following is not relevant for seed village with respect to its disadvantages?

   (A) Lack of co-operation among farmers
   (B) Engage trained labourers
   (C) Outbreak of diseases in whole area
   (D) Operational difficulties at a particular time

ANSWER SHEET

1. **D** 2. **C** 3. **A** 4. **B** 5. **A** 6. **D** 7. **B**

Chapter 9

PRINCIPLES OF SEED PRODUCTION

1. Who from the following has suggested the steps for maintaining genetic purity?
   (A) Baskin and Delouche (1972)
   (B) Baskin and Baskin (1974)
   (C) Hartmann and Kester (1968)
   (D) Harrington and Woodstock (1962)

2. Which one of the following is not relevant with respect to maintenance of genetic contamination?
   (A) Providing recommended isolation
   (B) Conducting germination test distance
   (C) Conducting grow out test
   (D) Roguing of seed fields

3. Which one of the following is the cause for expression of off types?
   (A) Recessive gene expression
   (B) Dominant gene expression
   (C) Co-dominant gene expression
   (D) Cytoplasmic expression

4. What is the exact stage of removal of off types in seed field?
   (A) After pollination
   (B) During fertilization
   (C) Before pollination
   (D) After fertilization

5. In which one of the following crop, the blonded seed occurs?
   (A) Peas (B) Greengram
   (C) Cluster bean (D) Blackgram

6. ---------- is experienced with dimpled seed when it is exposed to high temperature and humidity at the time of maturity.
   (A) Peas (B) Greengram
   (C) Cluster bean (D) Blackgram

7. The off coloured seeds are to be discarded in the case of ---------- seeds.
   (A) Peas (B) Greengram
   (C) Cluster bean (D) Blackgram

8. When the seeds are exposed to high temperature at the time of development and maturation, the hard seeds are developed. In which crop, this is noticed.
   (A) Peas (B) Greengram
   (C) Cluster bean (D) Blackgram

9. Which one of the following is not suitable weather for quality seed production?
   (A) Moderate rainfall
   (B) Moderate wind
   (C) Optimum temperature
   (D) Humid climate

10. Which one of the following is not relevant for land selection for seed production?
   (A) Moderate water facilities
   (B) Good fertile soil
   (C) Neutral pH
   (D) Good drainage facilities

11. Which one of the following is not applicable for land selection for seed production?

(A) Avoid good drainage facilities

(B) Avoid coastal belts and marshy places

(C) Avoid areas where isolation is a problem

(D) Avoid areas of endemic diseases and pests

12. For which crop, the isolation distance is more or higher?

(A) Self pollinated crop

(B) Self fertilized crop

(C) Cross pollinated crop

(D) Often cross pollinated crop

13. Wind is the pollinating agent for

(A) Pearl millet (B) Red gram

(C) Castor (D) Sunflower

14. Bees are the pollinating agent for

(A) Pearl millet (B) Red gram

(C) Castor (D) Sunflower

15. For which crop, the isolation distance is less?

(A) Self pollinated crop

(B) Cross pollinated crop

(C) Cross fertilized crop

(D ... n cross pollinated crop

... rop is used for what type of

... Time isolation

(B) Distance isolation

(C) Barrier isolation

(D) Space isolation

17. Growing tomato seed crops in poorly drained soil cause which disease?

(A) Root rot (B) Root scortching

(C) Wilt (D) Damping off

18. Growing ----------- seed crop in poorly drained soil cause wilt disease

(A) Brinjal (B) Bhendi

(C) Pumpkin (D) Tomato

19. What is the symptom of potassium deficiency in tomato?

(A) Black rot

(B) Whip tail

(C) Blossom end rot

(D) Marsh spot

20. What is the symptom of boron deficiency in cabbage and cauliflower?

(A) Black rot

(B) Whip tail

(C) Blossom end rot

(D) Marsh spot

21. What is the symptom of molybdenum deficiency in crucifers?

(A) Black rot

(B) Whip tail

(C) Blossom end rot

(D) Marsh spot

22. What is the deficiency symptom of manganese in peas?

(A) Black rot (B) Whip tail

(C) Blossom end rot (D) Marsh spot

23. What is the deficiency symptom of boron in garden pea?

(A) Blossom end rot

(B) Hollow heart

(C) Darkened plumule

(D) Blotching

24. What is the symptom of calcium deficiency?

(A) Blossom end rot

(B) Hollow heart

(C) Darkened plumule

(D) Blotching

25. What is the symptom of calcium deficiency in tomato?
(A) Blossom end rot
(B) Hollow heart
(C) Darkened plumule
(D) Blotching

26. Which is not relevant for over or excess irrigation of seed crops?
(A) Promote vegetative growth
(B) Promote lodging
(C) Promote nutrient imbalance
(D) Delayed flowering

27. ---------- is noticed due to under irrigation
(A) Promote vegetative growth
(B) Promote lodging
(C) Promote nutrient imbalance
(D) Delayed flowering

28. Which one of the following weedicide is a post emergence weedicide?
(A) Atrazine (B) Glyphosate
(C) Butachlor (D) Pendimethalin

29. Which one of the following is the pre-emergence weedicide?
(A) Glyphosate (B) Atrazine
(C) 2,4-D (D) Paraquat

30. Which one of the following is sprayed to pulses for the production of bolder seeds?
(A) 0.5% DAP (B) 1% DAP
(C) 1.5% DAP (D) 2% DAP

31. In cotton, spraying of ---------- DAP at boll formation increases seed quality
(A) 0.5% (B) 1%
(C) 1.5% (D) 2%

32. Which from the following contaminant does not cause physical contamination?
(A) Pollen shedders
(B) Weed seeds
(C) Designated diseased plants
(D) Other crop plants

33. Which one of the following crop can be harvested in single harvest?
(A) Cotton (B) Tomato
(C) Rice (D) Chillies

34. Which one of the following crop can be harvested in multi picking?
(A) Rice (B) Maize
(C) Sorghum (D) Blackgram

35. Artificial drying is done at 30-32°C, but never exceeding ------------ °C
(A) 34°C (B) 36°C
(C) 38°C (D) 40°C

36. Which one of the following condition is ideal for storage?
(A) Wet and humid condition
(B) Dry and cool condition
(C) Dry and high temperature
(D) Dry and humid condition

37. Which one of the following is not relevant for achieving synchronized flowering?
(A) Staggered sowing
(B) Application of CCC
(C) Withholding irrigation
(D) Application of urea or DAP

38. Which one of the following is a genetic contaminant in cotton hybrid seed production
(A) Selfed bolls (B) Crossed bolls
(C) Skewed bolls (D) Unopened boll

39. Minerals essential for formation of s structures include
(A) Calcium (B) Carbon
(C) Cadmium (D) Phosphate

40. Rainfall, temperature, light, wind and humidity are
(A) Climatic factors
(B) Edaphic factors
(C) Biotic factors
(D) Topographic factors

ANSWER SHEET

1. **C**	2. **B**	3. **A**	4. **C**	5. **A**	6. **C**	7. **B**	8. **B & D**	9. **D**	10. **A**
11. **A**	12. **C**	13. **B**	14. **D**	15. **A**	16. **C**	17. **C**	18. **D**	19. **C**	20. **A**
21. **B**	22. **D**	23. **B**	24. **C**	25. **D**	26. **D**	27. **D**	28. **B**	29. **B**	30. **A**
31. **D**	32. **A**	33. **C**	34. **D**	35. **C**	36. **B**	37. **B**	38. **A**	39. **A**	40. **A**

Chapter 10

SEED PRODUCTION IN AGRICULTURAL CROPS

1. What is the isolation distance for wheat certified seed production?
   (A) 200 m (B) 150 m
   (C) 50 m (D) 3 m

2. What is the isolation distance for certified seed production of mustard?
   (A) 200 m (B) 150 m
   (C) 50 m (D) 3 m

3. What is the isolation distance for sunflower certified seed production?
   (A) 200 m (B) 150 m
   (C) 50 m (D) 3 m

4. What is the isolation distance for castor certified seed production?
   (A) 200 m (B) 150 m
   (C) 50 m (D) 3 m

5. ---------------- m is the isolation distance for foundation seed production of maize.
   (A) 400 (B) 10
   (C) 3 (D) 50

6. --------------- m is the isolation distance for foundation seed production of soybean.
   (A) 400 (B) 10
   (C) 3 (D) 50

7. -------------- m is the isolation distance for foundation seed production of jute.
   (A) 400 (B) 10
   (C) 3 (D) 50

8. --------------- m is the isolation distance for lab-lab foundation seed production.
   (A) 400 (B) 10
   (C) 3 (D) 50

9. Which one of the following is the kharif season?
   (A) February – May
   (B) June – September
   (C) October – January
   (D) November – December

10. Which from the following is the rabi season?
    (A) February – May
    (B) June – September
    (C) October – January
    (D) November – December

11. Which from the following is the summer season?
    (A) February – May
    (B) June – September
    (C) October – January
    (D) November – December

12. Wheat is sown in the month of -------------- for seed production.

(A) August – September
(B) April – August
(C) October – November
(D) November

13. The sowing season for certified seed production of sorghum is ----------------

(A) August – September
(B) April – August
(C) October – November
(D) November

14. For seed production of sunflower, the sowing season is ------------------

(A) August – September
(B) April – August
(C) October – November
(D) November

15. For seed production of cotton, the sowing season is

(A) August – September
(B) April – August
(C) October – November
(D) November

16. What is the seed rate for short duration rice?

(A) 10 kg/ha
(B) 60 kg/ha
(C) 100 - 130 kg/ha
(D) 35-40 kg/ha

17. --------------- is the seed rate for maize?

(A) 10 kg/ha
(B) 60 kg/ha
(C) 100 - 130 kg/ha
(D) 35-40 kg/ha

18. Which one of the following is the seed rate for bunchy type groundnut?

(A) 10 kg/ha
(B) 60 kg/ha
(C) 100 - 130 kg/ha
(D) 35-40 kg/ha

19. What is the seed rate for bengalgram?

(A) 10 kg/ha
(B) 60 kg/ha
(C) 100 - 130 kg/ha
(D) 35-40 kg/ha

20. Sorghum seeds can be soaked in ------------------ for 10h and dried back to original moisture content for enhancing germination and vigour under dryland conditions.

(A) 2% KCl (B) 0.5% $CaCl_2$
(C) 2% KH_2PO_4 (D) 100ml H_2SO_4

21. Which one of the following treatment is adopted for hardening of bajra seeds?

(A) 2% KCl (B) 0.5% $CaCl_2$
(C) 2% KH_2PO_4 (D) 100ml H_2SO_4

22. In groundnut, the seeds are soaked in ---------------- for 6h in equal volume of water for presprounting and separation of dead seeds.

(A) 2% KCl (B) 0.5% $CaCl_2$
(C) 2% KH_2PO_4 (D) 100ml H_2SO_4

23. Which from the following is used for delinting of cotton seeds.

(A) 2% KCl (B) 0.5% $CaCl_2$
(C) 2% KH_2PO_4 (D) 100ml H_2SO_4

24. What is the dosage of common salt used to dissolve in 10 litres of water for grading of paddy seeds using egg floatation technique?

(A) 0.5 kg (B) 1.0 kg
(C) 1.5 kg (D) 2.0 kg

25. For egg floatation technique to upgrade paddy seeds, 2kg of commercial common salt is to be dissolved in ----------- litres of water.

(A) 5 (B) 10
(C) 15 (D) 20

26. How many kilograms of paddy seeds are added to a salt solution in which 2 kg of commercial common salt is dissolved in 10 litres of water?

(A) 5 kg (B) 10 kg
(C) 15 kg (D) 20 kg

27. How to prepare 0.5% $CaCl_2$ for soaking of groundnut seeds?

(A) 125 g of $CaCl_2$ in 25 litres of water
(B) 125 g of $CaCl_2$ in 20 litres of water
(C) 125 g of $CaCl_2$ in 15 litres of water
(D) 125 g of $CaCl_2$ in 10 litres of water

28. After acid delinting and washing, which chemical is used to test the complete removal of acid

(A) Calcium oxychloride @ 0.5%
(B) Calcium bicarbonate @ 0.5%
(C) Calcium chloride @ 0.5%
(D) Calcium hydroxide @ 0.5%

29. A spacing of ---------------- is adopted for blackgram

(A) 90 x 45 cm (B) 30 x 30 cm
(C) 45 x 15 cm (D) 30 x 10 cm

30. A spacing of -------------- is adopted for gingelly

(A) 90 x 45 cm (B) 30 x 30 cm
(C) 45 x 15 cm (D) 30 x 10 cm

31. For seed roduction of cotton, the recommended spacing is ---------------

(A) 90 x 45 cm (B) 30 x 30 cm
(C) 45 x 15 cm (D) 30 x 10 cm

32. For seed production of sorghum, the recommended spacing is --------------

(A) 90 x 45 cm (B) 30 x 30 cm
(C) 45 x 15 cm (D) 30 x 10 cm

33. The percentage of offtypes permitted for certified seed production of wheat is -------

(A) 0.05% (B) 0.1%
(C) 0.2% (D) 0.5%

34. What is the permitted offtypes percentage for certified seed production of sorghum?

(A) 0.05% (B) 0.1%
(C) 0.2% (D) 0.5%

35. Which one of the following is offtypes permitted for soybean certified seed production?

(A) 0.05% (B) 0.1%
(C) 0.2% (D) 0.5%

36. The bunt is the designated disease in ------

(A) Sunflower (B) Rice
(C) Wheat (D) Cotton

37. Downy mildew is the designated disease in ---------------

(A) Sunflower (B) Rice
(C) Wheat (D) Cotton

38. Black arm of ------------------- is the designated disease

(A) Sunflower (B) Rice
(C) Wheat (D) Cotton

39. The designated disease of ---------------- is loose smut

(A) Sunflower (B) Rice
(C) Wheat (D) Cotton

40. Which one of the following is the objectionable weeds of wheat?

(A) Wild rice (B) Wild morning glory
(C) Wild Helianthus
(D) Wild oats

41. Which one of the following is the recommended sieve size for size grading of rice seed?

(A) 23/64" (B) 8/64"
(C) 24/64" (D) 27/64"

42. Which one of the following is the recommended sieve size for size grading of red gram seeds?

(A) 23/64" (B) 8/64"
(C) 24/64" (D) 27/64"

43. -------------- is the recommended sieve size for size grading of pearlmillet seed

(A) 23/64" (B) 8/64"
(C) 24/64" (D) 27/64"

44. -------------- is the recommended sieve size for grading of maize seed

(A) 23/64" (B) 8/64"
(C) 24/64" (D) 27/64"

45. Other crop seed permitted for black gram

(A) 5/kg (B) 10/kg
(C) 20/kg (D) 25/kg

46. Other crop seed permitted for cowpea

(A) 5/kg (B) 10/kg
(C) 20/kg (D) 25/kg

47. ----------- is the permitted other crop seed for wheat.

(A) 5/kg (B) 10/kg
(C) 20/kg (D) 25/kg

48. What is the permitted weed seed in redgram?

(A) 5/kg (B) 10/kg
(C) 20/kg (D) 40/kg

49. Which one of the following is permitted weed seed in rice?

(A) 5/kg (B) 10/kg
(C) 20/kg (D) 40/kg

50. What is the permitted weed seed in groundnut?

(A) 5/kg (B) 10/kg
(C) 20/kg (D) 40/kg

51. What is the permitted weed seed in sunflower?

(A) 5/kg (B) 10/kg
(C) 20/kg (D) 40/kg

52. The minimum standard of 65% germination is permitted for --------- seed.

(A) Wheat (B) Cotton
(C) Greengram (D) Castor

53. 85% minimum seed certification standard for germination is permitted for --------- seed.

(A) Wheat (B) Cotton
(C) Greengram (D) Castor

54. A germination of 70% is allowed

(A) Wheat (B) Cotton
(C) Greengram (D) Castor

55. For ------------------ seed, 75% germination is permitted as minimum standard

(A) Wheat (B) Cotton
(C) Greengram (D) Castor

56. In sunflower, production of foundation seed requires an isolation distance of (metres)

(A) 400 (B) 800
(C) 200 (D) 100

57. In wheat, production of foundation seed needs an isolation distance of

(A) Three metres (B) Five metres
(C) Ten metres (D) Twenty metres

58. Physical purity is permissible (98%) in the crop :

(A) Groundnut (B) Rice
(C) Soyabean (D) Spinach

59. Production of breeder seed in cotton requires an isolation distance of

(A) 20 metres (B) 30 metres
(C) 50 metres (D) 75 metres

60. What is seed multiplication ratio for potato and groundnut respectively?

(A) 1:4 and 1:8 (B) 1:8 and 1:4
(C) 1:8 and 1:10 (D) 1:10 and 1:8

61. Castor bean is

(A) Ornamental plant
(B) Cereal crop
(C) Pulse crop
(D) Oil seed crop

62. Which of the following is ploidy level of common wheat?
(A) Tetraploid (B) Hexaploid
(C) Diploid (D) Octaploid

63. The scientific name of bread wheat is
(A) *T. durum* (B) *T. dicoccum*
(C) *T. aestivum* (C) *T. monecoccum*

64. The scientific name of maize is
(A) *Glycine max* (B) *Hordeum vulgare*
(C) *Zea mays* (D) *Oryza sativa*

65. Triticale is a cross between
(A) Wheat & Barley
(B) Barley & Rye
(C) Wheat & Rye
(D) Wheat & Oat

66. Wheat crop is
(A) Cross pollinated
(B) Often cross pollinated
(C) Self pollinated
(D) Anemophily

67. Oat is said to be
(A) Cross pollinated crop
(B) Self pollinated crop
(C) Mixed pollinated crop
(D) Often cross pollinated crop

ANSWER SHEET

1. **D**	2. **C**	3. **A**	4. **B**	5. **A**	6. **C**	7. **D**	8. **B**	9. **B**	10. **C**
11. **A**	12. **D**	13. **C**	14. **B**	15. **A**	16. **B**	17. **A**	18. **C**	19. **D**	20. **C**
21. **A**	22. **B**	23. **D**	24. **D**	25. **B**	26. **B**	27. **A**	28. **C**	29. **D**	30. **B**
31. **A**	32. **C**	33. **C**	34. **B**	35. **D**	36. **B**	37. **A**	38. **D**	39. **C**	40. **B**
41. **A**	42. **C**	43. **B**	44. **D**	45. **A**	46. **B**	47. **C**	48. **D**	49. **C**	50. **B**
51. **A**	52. **B**	53. **A**	54. **D**	55. **C**	56. **A**	57. **A**	58. **B**	59. **C**	60. **A**
61. **D**	62. **B**	63. **C**	64. **C**	65. **C**	66. **C**	67. **B**			

Chapter 11

SEED PRODUCTION IN VEGETABLES

1. Among vegetables, which one of the following crop needs a less isolation distance

   (A) Brinjal (B) Crucifers
   (C) Tomato (D) Bhendi

2. What is the isolation distance for foundation seed production of tomato?

   (A) 50 m (B) 400 m
   (C) 1600 m` (D) 1000 m

3. -------------- is the isolation distance for FS of onion

   (A) 50 m (B) 400 m
   (C) 1600 m (D) 1000 m

4. Which one of the following is the isolation distance for FS of crucifers?

   (A) 50 m (B) 400 m
   (C) 1600 m (D) 1000 m

5. Which from the following is the isolation distance for FS of bhendi?

   (A) 50 m (B) 400 m
   (C) 1600 m (D) 1000 m

6. An isolation distance of 200 m is followed in certified seed production?

   (A) Brinjal (B) Cucurbits
   (C) Chilli (D) Onion mother bulb

7. Which one of the following vegetable, an isolation distance of 5m is followed in the case of certified seed production?

   (A) Brinjal (B) Cucurbits
   (C) Chilli (D) Onion mother bulb

8. The isolation distance of 500m is adopted for certified seed production of --------------

   (A) Brinjal (B) Cucurbits
   (C) Chilli (D) Onion mother bulb

9. For which vegetable an isolation distance of 100 m is adopted

   (A) Brinjal (B) Cucurbits
   (C) Chilli (D) Onion mother bulb

10. Which season is suitable for certified seed production of bittergourd

    (A) June – July
    (B) April – May
    (C) June – September
    (D) February - March

11. Which is a suitable season for certified seed production of tomato

    (A) June – July
    (B) April – May
    (C) June – September
    (D) February - March

12. --------------- is a suitable season for certified seed production of bhendi

(A) June – July
(B) April – May
(C) June – September
(D) September – October

13. ------------- is a suitable season for certified seed production of capsicum?

(A) June – July
(B) April – May
(C) June – September
(D) February - March

14. What is the seed rate for ashgourd?

(A) 300 – 400 g/ha
(B) 350 g/ha
(C) 2 kg/ha
(D) 1-1.5 kg/ha

15. What is the seed rate for chilli?

(A) 300 – 400 g/ha
(B) 350 g/ha
(C) 1 kg/ha
(D) 1-1.5 kg/ha

16. What is the seed rate for tomato?

(A) 300 – 400 g/ha
(B) 500 g/ha
(C) 1 kg/ha
(D) 1-1.5 kg/ha

17. ------------- is the seed rate for brinjal.

(A) 300 – 400 g/ha
(B) 500 g/ha
(C) 1 kg/ha
(D) 1-1.5 kg/ha

18. What seed treatment is recommended for cucurbits?

(A) Use of pregerminated seed for sowing
(B) Hydropriming (36 h)
(C) Hydropriming (12 h)
(D) Sand matric 80% priming (24 h)

19. --------------- is a best treatment for carrot?

(A) Use of pregerminated seed for sowing
(B) Hydropriming (36 h)
(C) Hydropriming (12 h)
(D) Sand matric 80% priming (24 h)

20. What seed treatment is recommended for onion?

(A) Use of pregerminated seed for sowing
(B) Hydropriming (36 h)
(C) Hydropriming (12 h)
(D) Sand matric 80% priming (24 h)

21. Which one of the following pretreatment is advocated for radish seed?

(A) Use of pregerminated seed for sowing
(B) Hydropriming (36 h)
(C) Hydropriming (12 h)
(D) Sand matric 80% priming (24 h)

22. A spacing of ---------------- is adopted for tomato.

(A) 45 x 30 cm (B) 60 x 60 cm
(C) 60 x 45 cm (D) 75 x 60 cm

23. A spacing of --------------- is adopted for brinjal.

(A) 45 x 30 cm (B) 60 x 60 cm
(C) 60 x 45 cm (D) 75 x 60 cm

24. What is the spacing for bhendi?

(A) 45 x 30 cm (B) 60 x 60 cm
(C) 60 x 45 cm (D) 75 x 60 cm

25. What is the spacing for chilli?

(A) 45 x 30 cm (B) 60 x 60 cm
(C) 60 x 45 cm (D) 75 x 60 cm

26. Which one of the following is a designated disease for muskmelon?

(A) Phompsis fruit blight
(B) Anthracnose
(C) Cucumber mosaic
(D) Black leg

27. ---------- is a designated disease for cabbage.

(A) Phompsis fruit blight
(B) Anthracnose
(C) Cucumber mosaic
(D) Black leg

28. Which from the following is a designated disease for chilli?

(A) Phompsis fruit blight
(B) Anthracnose
(C) Cucumber mosaic
(D) Black leg

29. -------------- is a designated disease for brinjal.

(A) Phompsis fruit blight
(B) Anthracnose
(C) Cucumber mosaic
(D) Black leg

30. Which one of the following is a objectionable weeds of bitter gourd?

(A) Balsam apple
(B) Weed melon
(C) Wild watermelon
(D) Jangali

31. ---------- is objectionable weeds of muskmelon

(A) Balsam apple
(B) Weed melon
(C) Wild watermelon
(D) Jangali

32. What is the objectionable weed of snakegourd?

(A) Balsam apple
(B) Weed melon
(C) Wild watermelon
(D) Jangali

33. Which one of the following is the optimum maturity stage of pumpkin?

(A) Yellow colouration of fruits
(B) Reddish colouration of fruits
(C) Ashy coating over the fruits
(D) Yellowish browning of fruit rind

34. ---------------------- is the optimum stage of harvesting of brinjal.

(A) Yellow colouration of fruits
(B) Reddish colouration of fruits
(C) Ashy coating over the fruits
(D) Yellowish browning of fruit rind

35. Which is the optimum stage of harvesting of ashgourd.

(A) Yellow colouration of fruits
(B) Reddish colouration of fruits
(C) Ashy coating over the fruits
(D) Yellowish browning of fruit rind

36. --------------- is the optimum stage of tomato fruits harvesting.

(A) Yellow colouration of fruits
(B) Reddish colouration of fruits
(C) Ashy coating over the fruits
(D) Yellowish browning of fruit rind

37. --------------- are recommended for tomato.

(A) 3-4 picking (B) 4-5 picking
(C) 6-8 picking (D) 8-9 picking

38. How many pickings are recommended for gourds?

(A) 3-4 picking (B) 4-5 picking
(C) 6-8 picking (D) 8-9 picking

39. Which one of the following method is best for tomato seed extraction?

(A) Fermentation method
(B) Dry seed extraction
(C) Acid method
(D) Alkali method

40. --------------------- is the seed extraction technique for cauliflower.

(A) Wet seed extraction
(B) Dry seed extraction
(C) Acid / Alkali method
(D) Fermentation method

41. What is the seed extraction method for okra?

(A) Wet seed extraction
(B) Fermentation seed extraction
(C) Acid / Alkali method
(D) Manual extraction

42. ---------------------- is the optimum sieve size for cucumber seed grading

(A) 5.75 / 64" (B) 9.75 / 64"
(C) 15/64" (D) 28/64"

43. Which one of the following is optimum sieve size for size grading of carrot seed?

(A) 5.75 / 64" (B) 9.75 / 64"
(C) 15/64" (D) 28/64"

44. -------------- is the optimum sieve size for size grading of watermelon seed

(A) 5.75 / 64" (B) 9.75 / 64"
(C) 15/64" (D) 28/64"

45. Which one of the following sieve size is used for size grading of bottlegourd seed?

(A) 5.75 / 64" (B) 9.75 / 64"
(C) 15/64" (D) 28/64"

46. ------------- is the optimum sieve size for size grading of onion seed

(A) 5.75 / 64" (B) 9.75 / 64"
(C) 15/64" (D) 28/64"

47. What is the pure seed percentage of okra?

(A) 95% (B) 97%
(C) 98% (D) 99%

48. ------------------- is the pure seed permitted for amaranthus.

(A) 95% (B) 97%
(C) 98% (D) 99%

49. Which one of the following is the pure seed permitted for tomato?

(A) 95% (B) 97%
(C) 98% (D) 99%

50. -------------- is the pure seed permitted for celery.

(A) 95% (B) 97%
(C) 98% (D) 99%

51. What is the number of other crop seeds allowed for certified seed of chilli?

(A) 5 kg / kg (B) 10 kg / kg
(C) 15 kg / kg (D) 20 kg / kg

52. -------------- is the other crop seeds permitted for certified seed of okra.

(A) 5 kg / kg (B) 10 kg / kg
(C) 15 kg / kg (D) 20 kg / kg

53. A minimum of 70% is the germination standard for

(A) Cauliflower (B) Carrot
(C) Cabbage (D) Curry leaf

54. For -------------, the minimum standard germination is 65%.

(A) Cauliflower (B) Carrot
(C) Cabbage (D) Curry leaf

55. A minimum of 60% is the germination standard for ---------------

(A) Cauliflower (B) Carrot
(C) Cabbage (D) Curry leaf

56. In bhendi, production of foundation seed needs an isolation distance of.

(A) 100 m (B) 50 m
(C) 200 m (D) 3 m

57. Physical purity of 95% is permissible for the foundation and certified seed of

(A) Soyabean
(B) Groundnut
(C) Spinach
(D) Carrot

58. Standards of germination for seed certification in chillies

(A) 70% (B) 90%
(C) 80% (D) 60%

59. Carrot, radish, cabbage and beet are
(A) Annual
(B) Biennial
(C) Perennial
(D) Woody perennial

60. Cauliflower, cabbage, broccoli, kohlrabi and Brussels sprouts have been derived from wild cabbage that, still grows in coastal regions of
(A) Latin America
(B) Mesopotamia
(C) Europe and N. Africa
(D) South Asia and South Africa

ANSWER SHEET

1. **C**	2. **A**	3. **D**	4. **C**	5. **B**	6. **C**	7. **D**	8. **B**	9. **A**	10. **A**
11. **C**	12. **A**	13. **B**	14. **D**	15. **C**	16. **A**	17. **A**	18. **A**	19. **B**	20. **D**
21. **C**	22. **D**	23. **C**	24. **A**	25. **B**	26. **C**	27. **D**	28. **B**	29. **A**	30. **A**
31. **B**	32. **D**	33. **D**	34. **A**	35. **C**	36. **B**	37. **C**	38. **B**	39. **C**	40. **B**
41. **D**	42. **A**	43. **A**	44. **C**	45. **D**	46. **B**	47. **D**	48. **A**	49. **C**	50. **A**
51. **B**	52. **A**	53. **C**	54. **A**	55. **B**	56. **C**	57. **D**	58. **D**	59. **B**	60. **C**

Chapter 12

TECHNIQUES OF HYBRID SEED PRODUCTION

1. Which one of the following is an example for two line hybrid seed production?
   (A) Sesamum (B) Jute
   (C) Groundnut (D) Cotton

2. Which from the following is an example for three line hybrid seed production?
   (A) Sorghum (B) Oats
   (C) Cotton (D) Tomato

3. An example for chemical hybridizing agent is
   (A) PEG (B) Mannitol
   (C) MH (D) Calcium chloride

4. Which one of the following is temperature dependent male sterility system?
   (A) PGMS (B) TGMS
   (C) LGMS (D) WGMS

5. Which one of the following is photopheriod dependent male sterility system?
   (A) PGMS (B) TGMS
   (C) LGMS (D) WGMS

6. Which from the following is intraspecific hybrid?
   (A) Varalakshmi (B) Sugana
   (C) Deccan (D) Ganga

7. Which from the following is interspecific hybrid?
   (A) Varalakshmi (B) Sugana
   (C) Deccan (D) Ganga

8. ---------------------- hybrid maize is an example for single cross hybrid
   (A) Jawahar (B) Deccan
   (C) COH1 (D) Ganga

9. ---------- hybrid maize is an example for double cross hybrid
   (A) Jawahar (B) Deccan
   (C) COH1 (D) Ganga

10. ------------- is an example for three way cross hybrid of maize
    (A) Jawahar (B) Deccan
    (C) COH1 (D) Ganga

11. In which of the following crop, the CMS is employed for hybrid seed production
    (A) Rice (B) Redgram
    (C) Maize (D) Cotton

12. Which from the following crop is an example for GMS method of hybrid seed production?
    (A) Rice (B) Redgram
    (C) Maize (D) Cotton

13. Detasseling and pollination is employed for hybrid seed production in -------------
    (A) Rice (B) Redgram
    (C) Maize (D) Cotton

14. Which from the following crop is an example for emasculation and dusting method of hybrid seed production?

(A) Rice (B) Redgram
(C) Maize (D) Cotton

15. Which method is used for hybrid seed production of tomato?

(A) Pinching of staminate flowers + hand pollination
(B) Hand emasculation + manual pollination
(C) GMS + bee pollination
(D) CMS + natural pollination

16. Which method is employed for bittergourd hybrid seed production?

(A) Pinching of staminate flowers + hand pollination
(B) Hand emasculation + manual pollination
(C) GMS + bee pollination
(D) CMS + natural pollination

17. Which from the following method is advocated for chilli hybrid seed production?

(A) Pinching of staminate flowers + hand pollination
(B) Hand emasculation + bee pollination
(C) GMS + bee pollination
(D) CMS + natural pollination

18. In onion hybrid seed production, ----------- method is employed

(A) Pinching of staminate flowers + hand pollination
(B) Hand emasculation + manual pollination
(C) GMS + bee pollination
(D) CMS + natural pollination

19. Which one of the following method is employed for watermelon hybrid seed production?

(A) Pinching of staminate flowers + hand pollination
(B) Removal of staminate flower + emasculation + hand pollination
(C) GMS + bee pollination
(D) CMS + natural pollination

20. What is the isolation distance for rice hybrid certified seed production?

(A) 100 m (B) 200 m
(C) 400 m (D) 200 m

21. What is the isolation distance for redgram hybrid certified seed production?

(A) 100 m (B) 200 m
(C) 400 m (D) 200 m

22. What is the isolation distance for sunflower hybrid certified seed production?

(A) 100 m (B) 200 m
(C) 400 m (D) 200 m

23. What is the isolation distance for onion hybrid certified seed production?

(A) 100 m (B) 200 m
(C) 400 m (D) 200 m

24. In which hybrid seed production, the time isolation is followed ?

(A) Sorghum (B) Pearlmillet
(C) Rice (D) Redgram

25. In -------------- hybrid seed production, the barrier isolation is used

(A) Sorghum (B) Pearlmillet
(C) Rice (D) Redgram

26. What is the seed rate for maize hybrid seed production?

(A) 20 kg female line & 10 kg male line/ha
(B) 12 kg female line & 5 kg male line/ha
(C) 16 kg female line & 9 kg male line/ha
(D) 4 kg female line & 1 kg male line/ha

27. Which one of the following is the seed rate for rice hybrid seed production?

(A) 20 kg female line & 10 kg male line/ha
(B) 12 kg female line & 5 kg male line/ha
(C) 16 kg female line & 9 kg male line/ha
(D) 4 kg female line & 1 kg male line/ha

28. Which from the following is the seed rate for bhendi hybrid seed production?

(A) 20 kg female line & 10 kg male line/ha
(B) 12 kg female line & 5 kg male line/ha
(C) 16 kg female line & 9 kg male line/ha
(D) 4 kg female line & 1 kg male line/ha

29. What is the seed rate for redgram hybrid seed production?

(A) 20 kg female line & 10 kg male line/ha
(B) 12 kg female line & 5 kg male line/ha
(C) 16 kg female line & 9 kg male line/ha
(D) 4 kg female line & 1 kg male line/ha

30. ---------- is the optimum planting ratio for sorghum hybrid seed production.

(A) 5:2 (B) 8:2
(C) 5:1 (D) 2:1

31. ----------- is the recommended planting ratio for cucumber hybrid seed production

(A) 5:2 (B) 8:2
(C) 5:1 (D) 2:1

32. Planting ratio for cotton hybrid seed production is

(A) 5:2 (B) 8:2
(C) 5:1 (D) 2:1

33. Planting ratio for tomato hybrid seed production is

(A) 5:2 (B) 8:2
(C) 5:1 (D) 2:1

34. At what time the hand pollination should be carried out to increase seed set in sunflower hybrid seed production?

(A) 5-8 am (B) 8-11 am
(C) 11 am -1 pm (D) 1– 4 pm

35. In which crop, rope pulling is done for effective pollination?

(A) Sunflower (B) Maize
(C) Rice (D) Bajra

36. ----------- % urea solution can be sprayed to late parent of sorghum hybrid seed production for enhancing flowering

(A) 1 (B) 2
(C) 3 (D) 4

37. In sorghum hybrid seed production, which of the following is sprayed to enhance flowering of late parent?

(A) 0.5% urea (B) 2% urea
(C) 1% urea (D) 300 ppm CCC

38. ------------ spray can be adopted for enhancing the flowering of parents in rice hybrid seed production.

(A) 0.5% urea (B) 2% urea
(C) 1% urea (D) 300 ppm CCC

39. Which one of the following spray can be used to delay flowering in sorghum seed production?

(A) 0.5% urea (B) 2% urea
(C) 1% urea (D) 300 ppm CCC

40. What is the dosage of GA_3 that is sprayed for enhancing flowering in rice hybrid seed production?

(A) 25 g/ha (B) 50 g/ha
(C) 75 g/ha (D) 100 g/ha

41. Which system is used for hybrid seed production in onion?

(A) Dioecious
(B) Cytoplasmic male sterility
(C) Genetic male sterility
(D) Self incompatibility

42. Pure line variety is a

(A) Homozygous – Heterogenous
(B) Homozygous – Homogenous
(C) Heterogenous – Heterogenous
(D) Heterogenous - Homozygous

43. Bagging is done to
(A) Avoid cross pollination
(B) Avoid self pollination
(C) Achieve desired pollination
(D) Prevent contamination from foreign pollen

44. The recommended ratio of male rows to female rows is in hybrid bajra production
(A) 2 to 4 (B) 1 to 2
(C) 4 to 8 (D) 3 to 6

45. The hybrids developed by Government Agencies or Government Institutions and Agricultural Universities are called
(A) Private hybrids
(B) Institutional hybrids
(C) Public hybrids
(D) Government

46. ------------------ deteriorates drastically upon inbreeding
(A) Carrot (B) Okra
(C) Brinjal (D) Broccoli

47. 'Experiments in Plant Hybridization' a paper of G.Mendal was first published in the year
(A) 1905 (B) 1787
(C) 1866 (D) 1896

48. 2n denotes :
(A) Number of homozygote produces by selfing a F1
(B) Number of types of gametes produced by F1
(C) Number of types of heterozygote produced
(D) Number of types of male gametes produced by F1

49. A cybrid is
(A) A sex hybrid
(B) A hybrid formed by cell fusion
(C) A plant produced by cell culture
(D) Symptom to hybrid

50. A homozygous line developed by self-fertilization in a cross-pollinated crop is termed as
(A) Pureline (B) Inbred
(C) Clone (D) Mutant

51. A hybrid is always
(A) Heterozygous
(B) Homozygous
(C) Often heterozygous
(D) Often homozygous

52. A maintainer line is
(A) Male fertile and can restore fertility
(B) Male sterile but can restore fertility
(C) Male sterile but cannot restore fertility
(D) Male fertile but cannot restore

53. A restorer gene is present in
(A) Nucleus (B) Cytoplasm
(C) Protoplsm (D) Golgi apparatus

54. Cross between cytoplasmic male sterile x cytoplasmic male fertile
(A) Produce sterile F1
(B) Produce fertile F1
(C) Produce fertile, sterile in 1:1 ratio
(D) Produce restorer F1

55. Emasculation is
(A) Removal of male parts from a bisexual flower
(B) Removal of female plants
(C) Removal of male and female parts
(D) Removal of ovary

56. Hybrid seeds can be produced through open pollination with the use of
(A) Inbreds (B) Clones
(C) Pure lines (D) Male sterile lines

57. In ---------------, pollen dispersal is poor and seed set is likely to be low on male sterile plants
(A) Tomato (B) Cauliflower
(C) Broccoli (D) Carrot

58. In bajra

(A) Cytoplasmic male sterility is found
(B) Genetic male sterility is found
(C) Cytoplasmic genetic male sterility is found
(D) Self-incompatibility is found

59. In cotton, the popularly grown commercial hybrid Varalaxmi is the product of

(A) Intraspecific hybridization in *Gossypium hirsutum*
(B) Intraspecific hybridization in *Gossypium barbadense*
(C) Interspecific hybridization between *Gossypium hirsutum* and *Gossypium barbadense*
(D) Interspecific hybridization between *Gossypium arboretum* and *Gossypium herbaceum*

60. Which sterility is exploited in hybrid seed production?

(A) Male genetic sterility
(B) Cytoplasmic sterility
(C) Cytoplasmic genetic male sterility
(D) Genetic sterility

61. Which one of the following chemicals is used to induce polyploidy?

(A) Ethyl methane sulphonate
(B) Methyl methane sulphonate
(C) Colchicines
(D) Nitrous acid

62. Which one of the following categories of seeds is the farmer expected to buy afresh every time?

(A) Variety (B) Hybrid
(C) Synthetic (D) Composite

63. Which of the following type(s) of male sterility is/are used in commercial seed production of double cross hybrid maize?

(A) Cytoplasm alone
(B) Genetic alone
(C) Cytoplasm and genetic
(D) Cytoplasmic and genetic with restors

64. Mendok is a gametocide used for inducing male sterility in

(A) Onion (B) Sunflower
(C) Maize (D) Cotton

65. Pseudo-fertility may also occur late in the season as in

(A) Tomato (B) Tobacco
(C) Tapioca (D) Tea

66. Pure line is

(A) The progeny of selfed heterozygous plant
(B) The progeny of selfed homozygous plant
(C) The progeny of vegetatively propagated
(D) The progeny of a nucleus seed plant

67. Restorer gene is needed

(A) To make a plant correct
(B) To bring fertility in male sterile plants
(C) To make a plant resistant
(D) To enhance the yield

68. Specific combining ability of the parents should be high in order to develop

(A) Hybrids (B) Composites
(C) Varieties (D) Synthetics

69. Sterility is most pronounced in the F1 generation following

(A) Intervarietal hybridization
(B) Intravarietal hybridization
(C) Intraspecific hybridization
(D) Interspecific hybridization

70. The initial pool of a composite population is composed of

(A) Inbred lines
(B) Isoline
(C) Pure lines
(D) Single plants derived from crosses and/or germplasm lines

71. The initial gene pool of a synthetic population is composed of

(A) Plants from an open-pollinated variety
(B) Pure lines
(C) Inbred lines
(D) Isolines

72. The male sterile line is a cross to produce hybrid seed is known as

(A) A Line (B) R line
(C) C line (D) None of them

73. The mechanism of male sterility in self-pollinated crops can be successfully utilised to obtain

(A) Selfed seed
(B) Mutated seed without radiation treatment
(C) Hybrid seed without emasculation
(D) Sterile seed

74. The type of male sterility used in hybrid seed production in pearlmillet is

(A) Genetic male sterility
(B) Cytoplasmic male sterility
(C) Genetic cytoplasmic male sterility
(D) None of the above

75. To produce hybrid seed we need

(A) A male sterile line, restorer, maintainer
(B) A male fertile line and maintainer
(C) A maintainer and restorer
(D) Only male sterile line

76. Which from the following is a F_1 hybrid?

(A) MSMS (B) msms
(C) MSms (D) MSMs

77. Which from the following is a pollen parent?

(A) MSms (B) MSMs
(C) MSMs (D) msms

78. Which from the following is a female parent?

(A) msms (B) MSMs
(C) MSMS (D) mSMs

79. At what temperature level, the TGMS line becomes sterile?

(A) > 34°C / 26°C (day / night)
(B) > 32°C / 24°C (day / night)
(C) < 24°C / 18°C (day / night)
(D) < 22°C / 16°C (day / night)

80. At what temperature level, TGSM line becomes fertile?

(A) > 34°C / 26°C (day / night)
(B) > 32°C / 24°C (day / night)
(C) < 24°C / 18°C (day / night)
(D) < 22°C / 16°C (day / night)

81. Which from the following gametocide is used for rice?

(A) Zinc methyl arsenate
(B) Maleic hydrazide
(C) FW450
(D) Gibberellin

82. Which from the following gametocide is used for wheat?

(A) Zinc methyl arsenate
(B) Maleic hydrazide
(C) FW450
(D) Gibberellin

83. Which from the following gametocide is used for cotton?

(A) Zinc methyl arsenate
(B) Maleic hydrazide
(C) FW450
(D) Gibberellin

84. Which from the following gametocide is used for sunflower?

(A) Zinc methyl arsenate
(B) Maleic hydrazide
(C) FW450
(D) Gibberellin

85. Which from the following gametocide is used for sugarbeet?
(A) Zinc methyl arsenate
(B) Maleic hydrazide
(C) FW450
(D) Chloroethyl phosphoric acid

86. Which from the following gametocide is used for rice?
(A) Ethrel
(B) Maleic hydrazide
(C) FW450
(D) Gibberellin

ANSWER SHEET

1. **D**	2. **A**	3. **C**	4. **B**	5. **A**	6. **B**	7. **A**	8. **C**	9. **B**	10. **D**
11. **A**	12. **B**	13. **C**	14. **D**	15. **B**	16. **A**	17. **C**	18. **D**	19. **B**	20. **A**
21. **B**	22. **C**	23. **D**	24. **C**	25. **C**	26. **B**	27. **A**	28. **D**	29. **C**	30. **A**
31. **D**	32. **B**	33. **C**	34. **B**	35. **C**	36. **A**	37. **C**	38. **B**	39. **D**	40. **C**
41. **B**	42. **B**	43. **D**	44. **A**	45. **C**	46. **A**	47. **C**	48. **B**	49. **B**	50. **B**
51. **A**	52. **D**	53. **A**	54. **A**	55. **A**	56. **A**	57. **A**	58. **C**	59. **C**	60. **C**
61. **C**	62. **B**	63. **D**	64. **A**	65. **B**	66. **B**	67. **B**	68. **A**	69. **D**	70. **C**
71. **C**	72. **A**	73. **C**	74. **C**	75. **A**	76. **C**	77. **B**	78. **A**	79. **B**	80. **C**
81. **A**	82. **B**	83. **C**	84. **D**	85. **D**	86. **A**				

Chapter 13

SEED PRODUCTION IN LEGUMES AND GRASSES

1. What is the isolation distance of certified seed of Lucerne?

   (A) 400 m (B) 300 m
   (C) 200 m (D) 100 m

2. The isolation distance for foundation seed production of stylo is

   (A) 25 m (B) 50 m
   (C) 75 m (D) 100 m

3. What is the seed rate of fodder cowpea?

   (A) 15 kg/ha
   (B) 10 kg/ha
   (C) 25 kg/ha
   (D) 20 kg/ha

4. The seed rate for Desmanthus seed production is

   (A) 15-20 kg/ha
   (B) 20-25 kg/ha
   (C) 10-20 kg/ha
   (D) 20 kg/ha

5. How many field inspections are done for cluster bean

   (A) 2 (B) 3
   (C) 4 (D) 5

6. Number of field inspections for stylo is

   (A) 2 (B) 3
   (C) 4 (D) 5

7. At what stage the Desmanthus seed crop is harvested

   (A) Green to dark brown
   (B) Green to light brown
   (C) Green to brown
   (D) Green to yellowish brown

8. How much quantity of seed sample is forwarded as working sample of berseem?

   (A) 4 g (B) 5 g
   (C) 6 g (D) 7 g

9. What is the minimum germination standard for Lucerne?

   (A) 70% (B) 75%
   (C) 80% (D) 85%

10. Which one of the following substrate is recommended for germination of fodder cowpea?

    (A) TP/S (B) BP/TP
    (C) TP/BP (D) BP/S

11. The isolation distance for foundation seed production of Cumbu-Napier hybrid

    (A) 10 m (B) 15 m
    (C) 20 m (D) 25 m

12. Seed rate of deenanath is ---------------- kg/ha

    (A) 1.5 (B) 2.0
    (C) 2.5 (D) 3.0

13. Number of field inspections for guinea grass is ------------

(A) 1 (B) 2
(C) 3 (D) 4

14. What is the germination medium for deenanath?

(A) Bp (B) Sand
(C) Soil (D) Tp

ANSWER SHEET

1. **D**	2. **B**	3. **C**	4. **B**	5. **A**	6. **B**	7. **A**	8. **C**	9. **C**	10. **D**
11. **A**	12. **C**	13. **C**	14. **D**						

Chapter 14

FLOWER SEED PRODUCTION

1. What is the botanical name of bird of paradise?

   (A) *Sterlitzia reginae*

   (B) *Dahlia imperiais*

   (C) *Solidago canadensin*

   (D) *Jagehus patula*

2. Which one of the following is the scientific name of marigold?

   (A) *Sterlitzia reginae*

   (B) *Dahlia imperiais*

   (C) *Solidago canadensin*

   (D) *Tagetes patula*

3. What is the origin of anthurium?

   (A) USA (B) UK

   (C) Mexico (D) Columbia

4. Origin of Orchid is -------------------

   (A) USA (B) UK

   (C) Southeast Asia (D) Columbia

5. What is the flower type of gladiolus?

   (A) Raceme (B) Spike

   (C) Head (D) Racemose

6. In China aster, the plants have both --------- and -----------

   (A) Imperfect flowers and staminate flowers

   (B) Perfect and pistillate flowers

   (C) Pistillate and staminate flowers

   (D) Male and female flowers

7. Flower of petunia is -------

   (A) Unisexual (B) Monoecious

   (C) Dioecious (D) Hermophrodite

8. In carnation, both -------- flowers are present

   (A) Unisexual, pistillate

   (B) Bisexual, staminate

   (C) Bisexual, occasionally unisexual

   (D) Unisexual, occasionally bisexual

9. In which month, chrysanthemum flowers?

   (A) April-May

   (B) February-March

   (C) January-February

   (D) June-July

10. Which type of mechanism favors cross pollination in anthurium?

    (A) Protandry

    (B) Protogyny

    (C) Self incompatible

    (D) Insects

11. In gladiolus, ---------------- aids in cross pollination

    (A) Protandry

    (B) Protogyny

    (C) Self incompatible

    (D) Insects

12. Propagation by seed is common in ---------

    (A) Petunia (B) Carnation

    (C) Gladiolus (D) Lilium

13. What is the optimum stage for harvest of seed in rose based on visual observation?

(A) Dried heads (B) Dried pods
(C) Dark purple (D) Orange, red stage

14. --------------- stage is recommended for harvesting of marigold for seed purpose

(A) Dried heads (B) Dried pods
(C) Dark purple (D) Orange, red stage

15. Storage behavior of orchid seed is ----------

(A) Recalcitrant (B) Orthodox
(C) Difficult (D) Intermediate

16. What is the optimum moisture for storage of marigold seed?

(A) 7% (B) 6%
(C) 9% (D) 8%

17. What is the botanical name of lotus?

(A) *Nelumbo nucifera*
(B) *Zinnia elegans*
(C) *Gypsophila elegans*
(D) *Nerium oleander*

18. Scientific name of Zinnia is ---------------

(A) *Zinnia elegans*
(B) *Zinnia globosa*
(C) *Zinnia japanica*
(D) *Zinnia reticulata*

19. Which one of the following is originated in Mexico?

(A) Cacti (B) Azalea
(C) Cosmos (D) Celosia

20. From which country, nerium is originated?

(A) Europe
(B) Mediterranean region
(C) America
(D) Mexico

21. What is the type of inflorescence in Bougainvillea?

(A) Solitary (B) Spike
(C) Panicle (D) Cluster

22. Inflorescence of heliconia is ----------------

(A) Solitary (B) Spike
(C) Panicle (D) Cluster

23. In gaillardia, ------------------ pollination is very common

(A) Cross (B) Self
(C) Often cross (D) Self and cross

24. What is the agent for pollination of heliconia?

(A) Water (B) Bird
(C) Wind (D) Bees

25. Seed propagation is the rule in ------------

(A) Cosmos (B) Cornflower
(C) Jasmine (D) Gypsophila

26. Visual symptom for harvesting of Celosia seed is -----------

(A) Black
(B) Brown
(C) Yellow
(D) Green to brown

27. Which one of the following flowering plant possess dormancy mechanism?

(A) Zinnia (B) Gaillardia
(C) Cactus (D) Cosmos

28. In Zinnia, the germination is improved by treating the seeds with ------------

(A) H_2O_2 (B) H_2SO_4
(C) Hcl (D) HNO_3

29. What is the isolation distance for CS production of Petunia hybrid?

(A) 200 m (B) 300 m
(C) 400 m (D) 500 m

30. For foundation seed production of Aster, an isolation distance of ---------- is ensured for quality seed production

(A) 300 m (B) 400 m
(C) 500 m (D) 600 m

31. What is the minimum seed germination standard for annual carnation as per ISTA rules?

(A) 75% (B) 85%
(C) 70% (D) 80%

32. For foundation seed production of ornamental sunflower, the minimum germination prescribed by seed testing laboratories is --------

(A) 75% (B) 65%
(C) 70% (D) 80%

33. A pure seed (minimum) of 97% is recommended for CS production of ---------

(A) Annual chrysanthemum
(B) Aster
(C) Ornamental sunflower
(D) Marigold

ANSWER SHEET

1. **A**	2. **D**	3. **D**	4. **C**	5. **B**	6. **B**	7. **D**	8. **C**	9. **A**	10. **B**
11. **A**	12. **A**	13. **C**	14. **A**	15. **B**	16. **D**	17. **A**	18. **A**	19. **C**	20. **B**
21. **D**	22. **C**	23. **A**	24. **B**	25. **D**	26. **A**	27. **C**	28. **A**	29. **C**	30. **B**
31. **A**	32. **C**	33. **D**							

Chapter 15

FRUIT SEED PRODUCTION

1. Mango belongs to which family?
   (A) Myrtaceae (B) Moraceae
   (C) Musaceae (D) Anacardiaceae

2. Type of inflorescence of jack fruit is-------
   (A) Panicle (B) Cyme
   (C) Cauliflorus (D) Solitary

3. What is the botanical name of date palm?
   (A) *Passiflora edulis*
   (B) *Phoenix dactylifera*
   (C) *Mangifera indica*
   (D) *Punica granatum*

4. Botanical name of apple is-----------------
   (A) *Malus domestica*
   (B) *Pyrus calleryana*
   (C) *Prunus cerasifera*
   (D) *Prunus armeniaca*

5. What is the type of inflorescence of almond
   (A) Solitary (B) Pancile
   (C) Raceme (D) Corymbose

6. Which one of the following is the inflorescence of durian
   (A) Panicle (B) Corymbose
   (C) Raceme (D) Ramiflorous

7. The inflorescence of apple is
   (A) Cyme (B) Raceme
   (C) Panicle (D) Solitary

8. Which one of the following is the male inflorescence of walnut?
   (A) Panicle (B) Solitary
   (C) Raceme (D) Catkin

9. ----------------- is the inflorescence of avocado
   (A) Compound panicle
   (B) Compound raceme
   (C) Terminal panicle
   (D) Terminal raceme

10. ------------- pollination favours double fertilization in banana
    (A) Self (B) Cross
    (C) Often cross (D) Anthesis

11. Which one of the following is the pollinating agent in persimmon
    (A) Honey bees (B) Wind
    (C) Insects (D) Bat

12. Pomegranate flowers during---------------
    (A) January-February
    (B) March-April
    (C) June-July
    (D) September-October

13. Mango belongs to which type of fruit?
    (A) Berry (B) Aggregate fruit
    (C) Drupe (D) Hesperidium

14. Papaya is --------------
    (A) Dioecious (B) Monoecious
    (C) Hermophrodite (D) Bisexual

15. What is the type of fruit in jack?

(A) Berry (B) Aggregate fruit
(C) Drupe (D) Hesperidium

16. Mango is propagated by----------------

(A) Seed only
(B) Vegetative propagation
(C) Both seed and vegetative propagation
(D) Asexual only

17. Which one of the following is a recalcitrant?

(A) Banana (B) Papaya
(C) Guava (D) Mango

18. Jack fruit seed is very short lived because it is ------------------

(A) Orthodox seed
(B) Difficult recalcitrant seed
(C) Recalcitrant seed
(D) Intermediate seed

19. For storage, the safer moisture for citrus is -------------------

(A) 35-40% (B) 25-30%
(C) 10-15% (D) 10-15%

20. Persimmon flower belongs to --------------

(A) Monoecious
(B) Dioecious
(C) Andromonoecious
(D) Gynomonoecious

21. Which type of fruit in Avocado

(A) Drupe (B) Synconium
(C) Berry (D) Capsule

22. The fruit of West Indian cherry is ---------

(A) Drupe (B) Synconium
(C) Berry (D) Capsule

23. Which one of the following is the flower type of pomegranate

(A) Monoecious (B) Dioecious
(C) Andromonoecious
(D) Gynomonoecious

24. What type of pollination in date palm

(A) Entomophily
(B) Chiropterophily
(C) Self pollination
(D) Often cross pollination

25. Which type of propagation in mangosteen?

(A) Seed
(B) Vegetative propagules
(C) Root stock
(D) Scion

26. Which of the following species does not possess dormancy?

(A) Loquat (B) Custard apple
(C) Pineapple (D) Litchi

27. In how many days, the developing seeds attains physiological maturity in date palm?

(A) 100 DAA (B)150 DAA
(C) 200 DAA (D) 250 DAA

28. Rambutan seeds attain maturity in --------- WAA

(A) 14-17 (B) 15-18
(C) 19-22 (D) 23-26

29. What type of dormancy exist in fig?

(A) Endogenous dormancy
(B) Chemical dormancy
(C) Seed coat dormancy
(D) Embryo dormancy

30. The -------------- dormancy is noticed in passion fruit.

(A) Endogenous dormancy
(B) Chemical dormancy
(C) Seed coat dormancy
(D) Embryo dormancy

31. What is the critical moisture for storage of avocado?

(A) 32% (B) 40%
(C) 15% (D) 10%

32. What is the critical moisture for storage of rambutan

(A) 32% (B) 40%
(C) 15% (D) 10%

33. What is the family of apple?

(A) Anacardiaceae
(B) Actinidiaceae
(C) Juglandaceae
(D) Rosaceae

34. Which one of the following is the family of Walnut?

(A) Anacardiaceae
(B) Actinidiaceae
(C) Juglandaceae
(D) Rosaceae

35. What is the fruit type of almond?

(A) Berry (B) Drupe
(C) Pome (D) Aggregate berry

36. What is the type of fruit of Kiwi?

(A) Berry (B) Drupe
(C) Pome (D) Aggregate berry

37. What is the time of anthesis ------------?

(A) 9-10 am (B) 10-11 am
(C) 11 am-12 pm (D) 8-9 am

38. Which one of the following is the pollination of apricot?

(A) Self pollination
(B) Often cross pollination
(C) Cross pollination
(D) Self pollination and cross pollination

39. Which one of the following is an agent to aid cross pollination in peach?

(A) Wind (B) Insect
(C) Water (D) Bees

40. Walnut is propagated by

(A) Leaf (B) Root
(C) Bud (D) Seed

41. Seed dormancy is commonly existing in --------------- seeds

(A) Avocado (B) Apple
(C) Litchi (D) West Indian cherry

42. Apple possess -------------- dormancy

(A) Seed coat (B) Chemical
(C) Secondary (D) Embryo

43. Which type of dormancy exist in pecan nut?

(A) Seed coat (B) Chemical
(C) Physical (D) Physiological

44. Which one of the following method is adopted to break dormancy of peach?

(A) Pre-chill @ 3°C (B) Pre-chil @ 4°C
(C) Pre-chill @ 5°C (D) Pre-chill @ 6°C

45. How to break the embryo dormancy of apple?

(A) Removal of fruit coat
(B) Removal of seed coat
(C) Removal of cotyledon
(D) Removal of inner and outer coats of seed

46. Which one of the following fruit seed is an orthodox in nature?

(A) Mango (B) Apple
(C) Jack (D) Citrus

47. For long term storage of pear, what is your recommendation?

(A) Temp 5°C, RH 10%
(B) Temp 5°C, RH 15%
(C) Temp 10°C, RH 10%
(D) Temp 10°C, RH 15%

48. What is the safer temperature and RH for storage of pecan nut?

(A) Temp. 10°C, 90% RH
(B) Temp. 5°C, 90% RH
(C) Temp. 5°C, 80% RH
(D) Temp. 5°C, 70% RH

49. What is the isolation distance for FS seed production of papaya?

(A) 100 m (B) 1000 m
(C) 200 m (D) 1500 m

50. Which one of the following is the pure live clones of CS of mango?

(A) 100% (B) 99%
(C) 98% (D) 97%

51. The pure live clones for FS storage of acid lime is --------------- %

(A) 99.5 (B) 98.5
(C) 97.5 (D) 96.5

52. What is the pure live clones percentage for CS storage of apple?

(A) 99.5 % (B) 98.5%
(C) 97.5 % (D) 96.5%

53. The other living plants including rootstock for CS of guava is

(A) 1% (B) 2%
(C) 3% (D) 4%

54. Which one of the following is the other living plants for FS of pear?

(A) 0.1% (B) 0.4%
(C) 0.6% (D) 0.5%

ANSWER SHEET

1. **D**	2. **C**	3. **B**	4. **A**	5. **A**	6. **D**	7. **A**	8. **D**	9. **A**	10. **B**
11. **A**	12. **B**	13. **C**	14. **A**	15. **B**	16. **C**	17. **D**	18. **C**	19. **A**	20. **B**
21. **C**	22. **A**	23. **C**	24. **A**	25. **A**	26. **D**	27. **C**	28. **B**	29. **C**	30. **C**
31. **A**	32. **A**	33. **D**	34. **C**	35. **B**	36. **A**	37. **B**	38. **D**	39. **A**	40. **D**
41. **B**	42. **D**	43. **C**	44. **A**	45. **D**	46. **B**	47. **A**	48. **B**	49. **D**	50. **C**
51. **A**	52. **B**	53. **B**	54. **D**						

Chapter 16

NURSERY MANAGEMENT

1. What is the height of raised bed?

(A) 5-10 cm
(B) 10-15 cm
(C) 15-20 cm
(D) 20-25 cm

2. How much DAP is mixed in the nursery area of 8 m x 0.6 m?

(A) 100 g/bed
(B) 200 g/bed
(C) 300 g/bed
(D) 400 g/bed

3. In the nursery, the seeds are sown ----------- apart in the furrow.

(A) 0.5 – 1.0 cm
(B) 1.0 – 1.5 cm
(C) 1.5 – 2.0 cm
(D) 2.0 – 2.5 cm

4. How much FYM is added in the nursery area of 8 m x 0.6 m?

(A) 50 -100 kg/bed
(B) 100 – 150 kg/ha
(C) 150-200 kg/ha
(D) 200-250 kg/ha

5. How much carbendazim 50% WP is recommended in the nursery area of 8m x 0.6m?

(A) 10 g/bed
(B) 15 g/bed
(C) 30 g/bed
(D) 20 g/bed

ANSWER SHEET

1. **B** 2. **B** 3. **D** 4. **C** 5. **C**

Chapter 17

MICROPROPAGATION

1. Which one of the following is not relevant with respect to micropropagation?
   (A) Production of virus free stock
   (B) To multiply plants whose multiplication rate is too high.
   (C) To obtain genetic variability
   (D) Germplasm conservation

2. A ----------------- is a group of undifferentiated plant cells which can undergo divisions to form all types of tissues.
   (A) Meristem (B) Callus
   (C) Embryo (D) Protoplast

3. Which one of the following is not related to embryo culture?
   (A) Used in propagation of haploid plants
   (B) Embryos are used as explants
   (C) Both immature and mature embryos are used
   (D) To generate callus cultures or somatic embryos

4. Which one of the following disinfectant is used for surface sterilization of explants?
   (A) Sodium chloride
   (B) Sodium bicarbonate
   (C) Sodium hypochlorite
   (D) Sodium hydroxide

5. In tissue culture, which hormone is used for inducing shoot bud formation.
   (A) GA (B) Auxins
   (C) ABA (D) Cytokinins

6. Which one of the following hormone is used for root formation?
   (A) GA (B) Auxins
   (C) ABA (D) Cytokinins

7. Which one of the following is used as a explant for banana tissue culture?
   (A) Suckers
   (B) Shoot tip
   (C) Meristem tip
   (D) Vegetative bud

8. ----------------- is a explants for Eucalyptus tissue culture.
   (A) Suckers (B) Shoot tip
   (C) Meristem tip (D) Vegetative bud

9. For carnation tissue culture, the explants is -----------------
   (A) Suckers (B) Shoot tip
   (C) Meristem tip (D) Vegetative bud

10. What is the explants for anthurium tissue culture?
    (A) Suckers (B) Shoot tip
    (C) Meristem tip (D) Vegetative bud

11. Which medium is followed in rose tissue culture?
    (A) Skoogs medium
    (B) Anderson formula
    (C) MS
    (D) WPM

12. Which one of the following is a medium for Eucalyptus tissue culture?

(A) Skoogs medium
(B) Anderson formula
(C) MS
(D) WPM

13. --------------- is a suitable medium for Rhododenron tissue culture.

(A) Skoogs medium
(B) Anderson formula
(C) MS
(D) WPM

ANSWER SHEET

1. **B** 2. **B** 3. **D** 4. **C** 5. **C** 6. **B** 7. **A** 8. **B** 9. **C** 10. **D**
11. **C** 12. **D** 13. **B**

Unit 3
POST HARVEST SEED MANAGEMENT

Chapter 1

HARVESTING AND THRESHING

1. In which of the following crops, harvesting is done once?

(A) Tomato (B) Cotton
(C) Groundnut (D) Redgram

2. In which of the following crops, the fruits are harvested in 5-6 pickings?

(A) Tomato (B) Cotton
(C) Gourds (D) Brinjal

3. -----------------, the fruits are harvested in 4-5 picks.

(A) Tomato (B) Cotton
(C) Gourds (D) Brinjal

4. In which of the following crop, machine harvesting can be done.

(A) Tomato (B) Brinjal
(C) Gourds (D) Rice

5. Harvesting of --------------------- is done by rubbing and squeezing.

(A) Rice (B) Greengram
(C) Groundnut (D) Tomato

6. The pods of ------------------- is removed by stripping.

(A) Rice (B) Greengram
(C) Groundnut (D) Tomato

7. The seeds of ------------- is extracted by beating.

(A) Rice (B) Greengram
(C) Groundnut (D) Tomato

8. Which one of the following is not a function of threshers?

(A) To grade the seed and deliver in the outlet
(B) To clear the seed and deliver in the outlet
(C) To thresh the seed out of the head
(D) To thresh the seed from the straw

9. What is the recovery percentage of tomato seeds extracted using acid method?

(A) 0.6 - 1.0%
(B) 0.8 – 1.2%
(C) 0.9 – 1.2%
(D) 1.0 – 1.2%

10. What is the recovery percentage of tomato seeds extracted by fermentation method?

(A) 0.6 - 1.0%
(B) 0.8 – 1.2%
(C) 0.9 – 1.2%
(D) 1.0 – 1.2%

11. Which one of the following is the washing soda that is used in alkali method of seed extraction?

(A) Sodium carbonate
(B) Sodium bicarbonate
(C) Sodium chloride
(D) Sodium hydroxide

12. Which one of the following chemical is used for seed extraction of ash gourd, but it is deleterious to germination and further storage?

(A) Sodium carbonate
(B) Sodium bicarbonate
(C) Sodium chloride
(D) Sodium hydroxide

13. Which one of the following is used for seed extraction of ash gourd?

(A) 1:0.4 cm HCl for ½ h
(B) 1:0.5 conc. HCl for ½ h
(C) 1:0.6 Conc. HCl for ½ h
(D) 1:0.7 Conc. HCl for ½ h

14. What is the seed recovery percentage in HCl method of extraction of ash gourd?

(A) 4-5% (B) 5-6%
(C) 6-7% (D) 7-8%

15. --------------- is the damage that is observed in machine extraction of ash gourd seed.

(A) 3-4.5% (B) 4-5.5%
(C) 5-6.5% (D) 6-7.5%

ANSWER SHEET

1. **C**	2. **B**	3. **C**	4. **D**	5. **A**	6. **C**	7. **B**	8. **A**	9. **C**	10. **A**
11. **A**	12. **D**	13. **C**	14. **B**	15. **A**					

Chapter 2

SEED DRYING

1. Elimination of moisture from the seed depends on -------------------- and ----------- of the environment surrounding the seed

   (A) RH and gases
   (B) RH and light
   (C) RH and temperature
   (D) RH and rainfall

2. Which one of the following is not connected with factors affecting seed drying

   (A) The rate of seed flow
   (B) The rate of air blow
   (C) Drying temperature
   (D) Static pressure

3. While drying, normally the moisture content of seeds is reduced to -------------

   (A) 14-16% (B) 13-15%
   (C) 12-14% (D) 10-12%

4. Which kind of seeds, LSU driers are used?

   (A) Redgram (B) Paddy
   (C) Blackgram (D) Bajra

5. For drying of paddy seeds, the suitable drying method is -----------------

   (A) Non mixing column driers
   (B) Rectangular metal bin driers
   (C) Circular metal bin driers
   (D) Square metal bin driers

6. The seeds having initial moisture content of 10-18% should be dried at a temperature of

   (A) 110°F (43.3°C)
   (B) 100°F (42.2°C)
   (C) 90°F (42.2°C)
   (D) 85°F (30.0°C)

7. At what temperature, the seeds possessing an initial moisture of 18-30% should be dried?

   (A) 110°F (43.3°C)
   (B) 100°F (42.2°C)
   (C) 90°F (42.2°C)
   (D) 85°F (30.0°C)

8. At what temperature, the seeds possessing an initial moisture content of 10% should be dried?

   (A) 110°F (43.3°C)
   (B) 100°F (42.2°C)
   (C) 90°F (42.2°C)
   (D) 85°F (30.0°C)

ANSWER SHEET

1. **C** 2. **A** 3. **D** 4. **B** 5. **A** 6. **B** 7. **C** 8. **A**

Chapter 3

SEED PROCESSING

1. Which one of the following is not relevant to the purpose of seed processing?

   (A) To remove the undesirable materials

   (B) To remove good seeds

   (C) To remove the weed seeds

   (D) To size grade the seeds

2. Which one of the following machine is used to upgrade the seed based on seed length?

   (A) Air screen cleaner cum grader

   (B) Indented cylinder separator

   (C) Specific gravity separator

   (D) Magnetic separator

3. ---------------- is used for size grading of seeds.

   (A) Air screen cleaner cum grader

   (B) Indented cylinder separator

   (C) Specific gravity separator

   (D) Magnetic separator

4. Which from the following machine is engaged for upgrading the seed based on weight?

   (A) Air screen cleaner cum grader

   (B) Indented cylinder separator

   (C) Specific gravity separator

   (D) Magnetic separator

5. ---------------- is engaged for upgrading the seeds based on seed surface texture.

   (A) Air screen cleaner cum grader

   (B) Indented cylinder separator

   (C) Specific gravity separator

   (D) Magnetic separator

ANSWER SHEET

1. **B** 2. **B** 3. **A** 4. **C** 5. **D**

Chapter 4

SEED QUALITY ENHANCEMENT

1. Which of the following seed fortification treatment is adopted for paddy.

   (A) 1% $CaCl_2$
   (B) 2% KCl
   (C) 0.5% Potassium dihydrogen chloride
   (D) 100 ppm $ZnSO_4$

2. For blackgram seed fortification, ------------ is advocated

   (A) 1% $CaCl_2$
   (B) 2% KCl
   (C) 0.5% Potassium dihydrogen chloride
   (D) 100 ppm $ZnSO_4$

3. Soaking of sorghum seed in ------------------ is recommended for fortification

   (A) 1% $CaCl_2$
   (B) 2% KCl
   (C) 0.5% Potassium dihydrogen chloride
   (D) 100 ppm $ZnSO_4$

4. Which one of the following treatment is used for fortification of groundnut

   (A) 1% $CaCl_2$
   (B) 2% KCl
   (C) 0.5% Potassium dihydrogen chloride
   (D) 100 ppm $ZnSO_4$

5. Which one of following organic solvent is used for seed infusion?

   (A) Ketone (B) Acetone
   (C) Acetyl CoA (D) Acetic acid

6. What is the chemical used for sorghum seed hardening?

   (A) 0.5% potassium dihydrogen phosphate
   (B) 1% potassium chloride
   (C) 100 ppm zinc sulphate
   (D) 100 ppm manganese sulphate

7. ------------------ is the best treatment for hardening of greengram seeds.

   (A) 0.5% potassium dihydrogen phosphate
   (B) 1% potassium chloride
   (C) 100 ppm zinc sulphate
   (D) 100 ppm manganese sulphate

8. Which one of the following is recommended for seed hardening of redgram.

   (A) 0.5% potassium dihydrogen phosphate
   (B) 1% potassium chloride
   (C) 100 ppm zinc sulphate
   (D) 100 ppm manganese sulphate

9. --------------- is recommended for hardening of finger millet seeds.

   (A) 0.5% potassium dihydrogen phosphate
   (B) 1% potassium chloride
   (C) 100 ppm zinc sulphate
   (D) 100 ppm manganese sulphate

10. What do you mean by halopriming?

    (A) Soaking in water
    (B) Soaking in osmotic solution
    (C) Soaking in salt solution
    (D) Soaking in biocontrol agents

11. Seed soaking in ---------------- is called as hydropriming.

(A) Water (B) KCl
(C) Pseudomonas (D) PEG

12. Seed soaking in --------------- is called as biopriming

(A) Water (B) KCl
(C) Pseudomonas (D) PEG

13. Seed soaking in ---------------- is called as halopriming

(A) Water (B) KCl
(C) Pseudomonas (D) PEG

14. Soaking of seeds in -------------------- is called as osmo priming

(A) Water (B) KCl
(C) Pseudomonas (D) PEG

15. What is the suggested temperature during seed priming?

(A) 0-5°C (B) 5-10°C
(C) 10-15°C (D) 15-20°C

ANSWER SHEET

1. **C**	2. **D**	3. **B**	4. **A**	5. **B**	6. **A**	7. **D**	8. **C**	9. **B**	10. **C**
11. **A**	12. **C**	13. **B**	14. **D**	15. **C**					

Unit 4
SEED LEGISLATION AND CERTIFICATION

Chapter 1

SEED ACT AND RULES

1. In which year, the Indian Seed Act was enacted?

   (A) 1946 (B) 1956
   (C) 1966 (D) 1976

2. In which year, the Seed Rules were framed?

   (A) 1968. (B) 1958
   (C) 1978 (D) 1988

3. Seed Control Order ------------ was promulgated.

   (A) 1973 (B) 1983
   (C) 1953 (D) 1963

4. When the Seed Act was implemented?

   (A) 1959 (B) 1949
   (C) 1979 (D) 1969

5. Which one of the following is not related to the functions of Central Seed Committee.

   (A) To advise central and state governments
   (B) To certify seeds of any notified varieties
   (C) To recommend standards for seed certification
   (D) To recommend grow-out standards

6. Which one of the following is not applicable to the functions of State Seed Certification Agency?

   (A) To recommend standards for grow-out test
   (B) To certify seeds of any notified varieties
   (C) To take sample and inspect seed lots
   (D) To grant certificates

7. Which one of the following point is not suitable to the duties of Seed Analysts?

   (A) To test the seed sample received from Seed Certification Officer
   (B) To send the test results to the person concerned
   (C) To submit report to state government from time to time
   (D) To take seed samples of any notified variety from sale point

8. Which from the following statement not in accordance to the duties of Seed Inspectors

   (A) To take seed samples of any notified variety from a person who is selling seeds
   (B) To enter into the sale point where he suspects any offence
   (C) To send the test results to the person from whom he received seed sample
   (D) To examine any record or register found in the place where he suspects offence

9. Which of the following class of seed not under persue of seed certification authority.

   (A) Breeder seed
   (B) Foundation seed
   (C) Certified seed (D) All the above

10. Which class of seed directly controlled by the breeders

(A) Breeder seed
(B) Nucleus seed
(C) Foundation seed
(D) Certified seed

11. Which of the following is not in line with the Seed Control Order, 1983

(A) Issue of licence to dealers
(B) Issue of test report to person
(C) Renewal of licence
(D) Appointing of licencing authority

12. How much fees is collected for renewal of licence under Seed Control Order, 1983

(A) Rs.25 (B) Rs.50
(C) Rs.75 (D) Rs.100

13. Which one of the following statement is not applicable for New Seed Policy, 1988.

(A) Import of quality seed
(B) Export of quality seed
(C) Modernize plant quarantine
(D) Incentives to encourage domestic seed industry

ANSWER SHEET

1. **C**	2. **A**	3. **B**	4. **D**	5. **B**	6. **A**	7. **D**	8. **C**	9. **A**	10. **A&B**
11. **B**	12. **A**	13. **B**							

Chapter 2

SEED CERTIFICATION

1. Which one of the following is not applicable to phases of seed certification?
   (A) Verification of seed source
   (B) Undertaking field inspection
   (C) Taking seed sample from processing unit
   (D) Taking seed sample from sale point

2. ---------------- is the maximum area that can be offered for certification for which a single application is submitted.
   (A) 25 acres (B) 20 acres
   (C) 15 acres (D) 10 acres

3. FORM-I for seed certification should be submitted within ---------------------- from the date of sowing.
   (A) 60 days (B) 50 days
   (C) 40 days (D) 30 days

4. FORM-I for seed certification should be submitted within ---------------------- from the date of transplanting.
   (A) 5 days (B) 10 days
   (C) 15 days (D) 20 days

5. How much time is allowed for the seeds to be certified from the date of harvest?
   (A) 1 ½ months (B) 2 ½ months
   (C) 3 ½ months (D) 4 ½ months

6. At which stage the pollen shedders are removed in the case of bajra hybrid seed production fields?
   (A) Preflowering phase
   (B) Flowering phase
   (C) Post flowering phase
   (D) Seed developmental phase

7. In which hybrid seed production, the pollen partials are removed and destroyed from seed production field?
   (A) Red gram (B) Cotton
   (C) Sorghum (D) Maize

8. Which one of the following hybrid seed production fields, the non removal of selfed bolls cause genetic contamination?
   (A) Red gram (B) Cotton
   (C) Sorghum (D) Maize

9. In ---------------- hybrid seed production fields, the fertile plants should be removed prior to anthesis for maintaining 100% genetic purity.
   (A) Red gram (B) Cotton
   (C) Sorghum (D) Maize

10. The shedding tassels are removed and destroyed during the entire flowering phase in order to maintain 100% genetic purity in the case of -------------- hybrid seed production fields.
    (A) Red gram (B) Cotton
    (C) Sorghum (D) Maize

11. How many counts to be taken by the Seed Certification Officer when the area of seed crop is 0-5 acres?
    (A) 5 counts (B) 6 counts
    (C) 7 counts (D) 8 counts

12. When the area of seed field is 16-20 acres, how many counts are taken by seed certification staff?

(A) 5 counts (B) 6 counts
(C) 7 counts (D) 8 counts

13. --------------- counts are taken by the Seed Certification Officer when the area sown for seed purpose is 6-10 acres.

(A) 5 counts (B) 6 counts
(C) 7 counts (D) 8 counts

14. If the area under seed crop is 11-15 acres, -------------- counts are taken for certification purpose.

(A) 5 counts (B) 6 counts
(C) 7 counts (D) 8 counts

15. How many counts are to be taken by the certification staff for seed crop that is grown in an area of 21-25 acres?

(A) 6 counts (B) 7 counts
(C) 8 counts (D) 9 counts

16. How many plants are to be observed for one count in the case of cotton seed crop?

(A) 50 plants (B) 100 plants
(C) 150 plants (D) 200 plants

17. In tomato seed production plot, the Seed Certification Officer is to observe ----------- plants per count.

(A) 50 plants (B) 100 plants
(C) 150 plants (D) 200 plants

18. In mustard seed production field, how many plants are to be observed for one count by the Seed Certification Staff?

(A) 500 plants (B) 400 plants
(C) 300 plants (D) 200 plants

19. How many number of plants are to be inspected by the Seed Certification Officer in wheat seed crop?

(A) 250 plants (B) 500 plants
(C) 750 plants (D) 1000 plants

20. The validity period for certified seed is ----

(A) 6 months (B) 9 months
(C) 12 months (D) 15 months

ANSWER SHEET

1. **D**	2. **A**	3. **D**	4. **C**	5. **B**	6. **B**	7. **C**	8. **B**	9. **A**	10. **D**
11. **A**	12. **D**	13. **B**	14. **C**	15. **D**	16. **B**	17. **B**	18. **A**	19. **D**	20. **B**

Chapter 3

FIELD INSPECTION

1. How many unit of certification has been considered when the area under seed production field is less than one hectare?

   (A) 1 unit (B) 2 units
   (C) 3 units (D) 4 units

2. If one field inspection requires 1000 plants/ ear heads and in one step one can observe 25 plants/earheads of seed crop. To complete one field count one has to take up how many steps?

   (A) 40 steps (B) 80 steps
   (C) 20 steps (D) 60 steps

3. In broadcasted seed crops like berseem and lucerne, how much area should be covered during field count?

   (A) 85% (B) 75%
   (C) 60-70% (D) 60%

4. How much area should be covered for field count in the case of non tillering cereals sown in line for example sorghum?

   (A) 85% (B) 75%
   (C) 60-70% (D) 60%

5. In line sowing wide spaced crops like cotton, ------------ area should be covered for field count

   (A) 85% (B) 75%
   (C) 60-70% (D) 60%

6. The area covered for field count is ---------- in the case of line sowing medium spaced seed crops like chickpea

   (A) 85% (B) 75%
   (C) 60-70% (D) 60%

7. How many field inspections can be undertaken by the Seed Certification Officer in groundnut seed crop?

   (A) 4 (B) 3
   (C) 2 (D) 1

8. In potato seed crop, the number of field inspections recommended is --------------

   (A) 4 (B) 3
   (C) 2 (D) 1

9. How many field inspections can be taken up for verification in okra seed crop

   (A) 4 (B) 3
   (C) 2 (D) 1

10. In radish seed crop, the number of field inspection recommend is ----------------

    (A) 4 (B) 3
    (C) 2 (D) 1

11. Which of the following seed crops required a minimum gap of one season while selecting the land for seed production?

    (A) Groundnut (B) Beerseem
    (C) Rice (D) Sesamum

12. Which of the following seed crops require a minimum gap of two seasons while selecting the land for seed production?

    (A) Groundnut (B) Beerseem
    (C) Rice (D) Sesamum

13. What is the minimum isolation distance required to avoid transmission of loose smut in the case of wheat?

(A) 50 m (B) 100 m
(C) 150 m (D) 200 m

14. What is the maximum permitted limit for off-types that is observed in guinea grass grown for seed production?

(A) 1.0% (B) 0.5%
(C) 0.2% (D) 0.1%

15. --------------- is the maximum permitted limit for off-types that is observed for rape seed grown for seed production.

(A) 1.0% (B) 0.5%
(C) 0.2% (D) 0.1%

16. In jute grown for certified seed production, the maximum permitted limit for off types is

(A) 1.0% (B) 0.5%
(C) 0.2% (D) 0.1%

17. What is the maximum permitted limit for off types that is observed in garlic grown for seed production?

(A) 1.0% (B) 0.5%
(C) 0.2% (D) 0.1%

ANSWER SHEET

1. **A**	2. **A**	3. **D**	4. **B**	5. **A**	6. **C**	7. **C**	8. **A**	9. **B**	10. **D**
11. **C&D**	12. **A&B**	13. **C**	14. **D**	15. **B**	16. **A**	17. **C**			

Chapter 4

OBJECTIONABLE WEEDS

1. What is the objectionable weed species in mustard?

(A) *Sorghum halepense*
(B) *Cuscuta* spp.
(C) *Cichorium oxycantha*
(D) *Argemone mexicana*

2. ----------------------- is the objectionable weed species in napier grass

(A) *Sorghum halepense*
(B) *Cuscuta* spp.
(C) *Cichorium oxycantha*
(D) *Argemone mexicana*

3. Which one of the following is a objectionable weed species of lucerne?

(A) *Sorghum halepense*
(B) *Cuscuta* spp.
(C) *Cichorium oxycantha*
(D) *Argemone mexicana*

4. What is the objectionable weed species of berseem?

(A) *Sorghum halepense*
(B) *Cuscuta* spp.
(C) *Cichorium oxycantha*
(D) *Argemone mexicana*

5. What is the maximum permitted limit of objectionable weeds at field level in the case of foundation seed production of rice?

(A) 0.01% (B) 0.02%
(C) 0.05% (D) 0.1%

6. Which one of the following is the maximum permitted limit of objectionable weeds in the case of foundation seed field of mustard?

(A) 0.01% (B) 0.02%
(C) 0.05% (D) 0.1%

7. ------------------ is the maximum permitted limit of objectionable weeds in certified seed production of lettuce.

(A) 0.01% (B) 0.02%
(C) 0.05% (D) 0.1%

8. Which one of the following is the maximum permitted limit of objectionable weeds in certified seed fields of rapeseed?

(A) 0.01% (B) 0.02%
(C) 0.05% (D) 0.1%

9. *Cuscuta* is

(A) Total stem parasite
(B) Partial stem parasite
(C) Partial root parasite
(D) Total root parasite

10. Plants that attach themselves to any neighbouring object often by means of some special devices and climb it are called

(A) Trailers
(B) Creepers
(C) Climbers
(D) Parasites

11. Plants that grown upon other plants but do not absorb food from them are called

(A) Parasites (B) Epiphytes
(C) Saprophytes (D) Symbionts

12. When two organisms live together for mutual help to each other are called

(A) Saprophytes (B) Symbionts
(C) Parasites (D) Epiphytes

13. Plants that grown upon other plants and absorb food from them are called :

(A) Parasites (B) Saprophytes
(C) Epiphytes (D) Symbionts

14. In India, parasitic weed of sorghum is:

(A) *Cuscuta* spp.
(B) *Striga* spp.
(C) *Orabanche* spp.
(D) *Loranthus* sp

ANSWER SHEET

1. **D**	2. **A**	3. **B**	4. **C**	5. **A**	6. **C**	7. **B**	8. **D**	9. **A**	10. **C**
11. **B**	12. **B**	13. **A**	14. **D**						

Chapter 5

SEED CERTIFICATION STANDARDS

1. Which one of the following is not relevant to field standards?
   (A) Recommended isolation distance
   (B) Minimum limit of offtypes
   (C) Minimum limit of pollen shedders
   (D) Maximum permissible limit of weed seeds

2. Which one of the following is not applicable for seed standards?
   (A) Maximum permissible limit of objectionable weed species
   (B) Minimum percentage of pure seeds
   (C) Maximum permissible limit of moisture content
   (D) Minimum germination percentage

3. What is the minimum moisture content of paddy seeds for certification purpose?
   (A) 7% (B) 8%
   (C) 12% (D) 13%

4. Which one of the following is the prescribed minimum standard of moisture content for soybean for certification?
   (A) 7% (B) 8%
   (C) 12% (D) 13%

5. What is the minimum moisture content of brinjal seeds for certification purpose?
   (A) 7% (B) 8%
   (C) 12% (D) 13%

6. ---------------- is the minimum moisture content of bittergourd seeds for certification.
   (A) 7% (B) 8%
   (C) 12% (D) 13%

7. Which one of the following is the minimum moisture content prescribed for peas for seed certification?
   (A) 7% (B) 8%
   (C) 9% (D) 10%

8. -------------------- is the minimum moisture content of cotton hybrids for seed certification.
   (A) 7% (B) 8%
   (C) 9% (D) 10%

9. What is the maximum pure seed prescribed for finger millet as a seed standard?
   (A) 97% (B) 98%
   (C) 99% (D) 100%

10. Which one of the following is the maximum standard for pure seed fraction for cowpea?
    (A) 97% (B) 98%
    (C) 99% (D) 100%

11. --------------- is the maximum standard for innert matter in the case of bhendi.
    (A) 1% (B) 2%
    (C) 3% (D) 4%

12. What is the maximum inert matter content for sesamum for issuing certification tags?
(A) 1% (B) 2%
(C) 3% (D) 4%

13. Which one of the following is the maximum inert matter of groundnut?
(A) 1% (B) 2%
(C) 3% (D) 4%

14. For maize, the maximum standard for inert matter is ---------------.
(A) 1% (B) 2%
(C) 3% (D) 4%

15. Which one of the following is the maximum standard permitted for other crop seeds for foundation seed of maize inbreds?
(A) 20/kg (B) 10/kg
(C) 5/kg (D) None

16. What is the maximum other crop seeds prescribed for certified seed of French bean?
(A) 20/kg (B) 10/kg
(C) 5/kg (D) None

17. What is the maximum standard for other crop seeds for certified seed of cluster bean?
(A) 20/kg (B) 10/kg
(C) 5/kg (D) None

18. ------------ is the maximum standard for other crop seed for foundation seed of sesamum.
(A) 20/kg (B) 10/kg
(C) 5/kg (D) None

19. What is the maximum standard for ODV for foundation seed of castor?
(A) None (B) 5/kg
(C) 10/kg (D) 20/kg

20. Which one of the following is the maximum standard for certified seed of cluster bean?
(A) None (B) 5/kg
(C) 10/kg (D) 20/kg

21. ------------ is the maximum standard for ODV for foundation seed of sorghum?
(A) None (B) 5/kg
(C) 10/kg (D) 20/kg

22. ------------ is the maximum standard for ODV for certified seed of safflower?
(A) None (B) 5/kg
(C) 10/kg (D) 20/kg

23. Which one of the following is the maximum standard permitted for OWS for certified seed of chilli?
(A) None (B) 5/kg
(C) 10/kg (D) 20/kg

24. -------------- is the maximum standard prescribed for OWS for certified seed of jute.
(A) None (B) 5/kg
(C) 10/kg (D) 20/kg

25. What is the maximum standard for OWS for foundation seed of sunflower?
(A) None (B) 5/kg
(C) 10/kg (D) 20/kg

26. For peas, the maximum limit for OWS is ------------------------.
(A) None (B) 5/kg
(C) 10/kg (D) 20/kg

27. What is the minimum standard for germination of certified seed of maize hybrid?
(A) 75% (B) 80%
(C) 85% (D) 90%

28. Which one of the following is the minimum standard for germination of foundation seed of redgram?
(A) 75% (B) 80%
(C) 85% (D) 90%

29. ---------------- is the minimum standard for germination of certified seed of wheat.

(A) 75% (B) 80%
(C) 85% (D) 90%

30. What is the minimum standard for germination of foundation seed of paddy?

(A) 75% (B) 80%
(C) 85% (D) 90%

31. ---------------- is the minimum standard for germination of certified seed of groundnut.

(A) 60% (B) 65%
(C) 70% (D) 75%

32. What is the minimum standard for germination of foundation seed of bottlegourd?

(A) 60% (B) 65%
(C) 70% (D) 75%

33. Which one of the following is the minimum germination standard for certified seed of bhendi?

(A) 60% (B) 65%
(C) 70% (D) 75%

34. ------------------ is the minimum standard for germination of certified seed of cotton hybrid.

(A) 60% (B) 65%
(C) 70% (D) 75%

ANSWER SHEET

1. **D**	2. **A**	3. **D**	4. **C**	5. **B**	6. **A**	7. **C**	8. **D**	9. **A**	10. **B**
11. **A**	12. **C**	13. **D**	14. **B**	15. **C**	16. **D**	17. **A**	18. **B**	19. **B**	20. **D**
21. **C**	22. **A**	23. **C**	24. **D**	25. **B**	26. **A**	27. **D**	28. **A**	29. **C**	30. **B**
31. **C**	32. **A**	33. **B**	34. **D**						

Chapter 6

SEED SAMPLING

1. A small portion of seed taken from different representative part of the seed lot is known as ------------------

(A) Primary sample
(B) Composite sample
(C) Submitted sample
(D) Working sample

2. A mixture of all the primary samples taken from the seed lot is named as ---------------

(A) Primary sample
(B) Composite sample
(C) Submitted sample
(D) Working sample

3. Which one of the following sample is used for purity analysis, germination test and other tests in the STL?

(A) Primary sample
(B) Composite sample
(C) Submitted sample
(D) Working sample

4. Working sample is the reduced representative of ------------------

(A) Primary sample
(B) Composite sample
(C) Submitted sample
(D) Working sample

5. What is the maximum weight of a seed lot for finger millet

(A) 40,000 kg (B) 20,000 kg
(C) 10,000 kg (D) 1,000 kg

6. ------------- is the maximum weight of a seed lot for maize.

(A) 40,000 kg (B) 20,000 kg
(C) 10,000 kg (D) 1,000 kg

7. What is the maximum weight of a seed lot for groundnut kernels

(A) 40,000 kg (B) 20,000 kg
(C) 10,000 kg (D) 1,000 kg

8. Which one of the following is the maximum weight of a seed lot for blackgram

(A) 40,000 kg (B) 20,000 kg
(C) 10,000 kg (D) 1,000 kg

9. Which one of the following is the maximum weight of a seed lot for sunnhemp

(A) 10,000 kg (B) 20,000 kg
(C) 30,000 kg (D) 40,000 kg

10. ------------- is the maximum weight of a seed lot for garlic

(A) 10,000 kg (B) 20,000 kg
(C) 30,000 kg (D) 40,000 kg

11. In sunflower, the maximum weight of a seed lot is --------------------

(A) 10,000 kg (B) 20,000 kg
(C) 30,000 kg (D) 40,000 kg

12. In lucerne, the maximum weight of a seed lot is ----------------------

(A) 40,000 kg (B) 30,000 kg
(C) 20,000 kg (D) 10,000 kg

13. What is the maximum weight of a seed lot of bhendi?

(A) 40,000 kg (B) 30,000 kg
(C) 20,000 kg (D) 10,000 kg

14. Which one of the following is the maximum weight of a seed lot of sweet potato

(A) 40,000 kg (B) 30,000 kg
(C) 20,000 kg (D) 10,000 kg

15. What is the maximum weight of a seed lot for daincha

(A) 30,000 kg (B) 10,000 kg
(C) 40,000 kg (D) 20,000 kg

16. In brinjal, the maximum weight of a seed lot is -------------------

(A) 30,000 kg (B) 10,000 kg
(C) 40,000 kg (D) 20,000 kg

17. ---------------- is the maximum weight of a seed lot for tapioca

(A) 30,000 kg (B) 10,000 kg
(C) 40,000 kg (D) 20,000 kg

18. In which container, the seed sample should be packed for testing moisture content at STL

(A) Moisture resistant container
(B) Moisture proof container
(C) Moisture vapour proof container
(D) Moisture pervious container

19. You are asked to take primary sample from a seed lot. If number of containers is upto 5, how many primary samples are drawn

(A) Atleast one in every 3 containers
(B) Atleast one in every 7 containers
(C) Atleast one in every 5 containers
(D) Atleast 5 primary samples

20. If the number of containers is >401, how many primary samples you should take

(A) Atleast one in every 3 containers
(B) Atleast one in every 7 containers
(C) Atleast one in every 5 containers
(D) Atleast 5 primary samples

21. How many primary samples are to be drawn from a seed lot having 6 to 30 containers?

(A) Atleast one in every 3 containers
(B) Atleast one in every 7 containers
(C) Atleast one in every 5 containers
(D) Atleast 5 primary samples

22. From -----, the primary samples are drawn, if a seed lot contains 31 to 400 containers

(A) Atleast one in every 3 containers
(B) Atleast one in every 7 containers
(C) Atleast one in every 5 containers
(D) Atleast 5 primary samples

23. Which one of the following size is the submitted sample of kidney bean

(A) 400 g (B) 500 g
(C) 750 g (D) 150 g

24. What is the submitted sample size of pearlmillet

(A) 400 g (B) 500 g
(C) 750 g (D) 150 g

25. ---------- is the submitted sample size of paddy

(A) 400 g (B) 500 g
(C) 750 g (D) 150 g

26. Which one of the following is the submitted sample size of horsegram

(A) 400 g (B) 500 g
(C) 750 g (D) 150 g

27. Which one of the following is the submitted sample size of niger

(A) 150 g (B) 160 g
(C) 250 g (D) 350 g

28. What is the submitted sample size of cotton varieties

(A) 150 g (B) 160 g
(C) 250 g (D) 350 g

29. ------------ is the submitted sample weight of mustard

(A) 150 g (B) 160 g
(C) 250 g (D) 350 g

30. What is the submitted sample weight of sunflower?

(A) 150 g (B) 160 g
(C) 250 g (D) 350 g

31. If the seed lot size ranges from 3001 to 20,000 kg, how many primary samples are drawn

(A) Atleast 5 individual samples
(B) One individual sample for each 300 kg, but not less than 5 samples
(C) One individual sample for each 500 kg, but not less than 10 samples
(D) One individual sample for each 700 kg, but not less than 40 samples

32. Which one of the following statement is correct, if you are asked to take samples in a seed lot size of 20,001 kg and above?

(A) Atleast 5 individual samples
(B) One individual sample for each 300 kg, but not less than 5 samples
(C) One individual sample for each 500 kg, but not less than 10 samples
(D) One individual sample for each 700 kg, but not less than 40 samples

33. How many number of primary samples are to be drawn from a seed lot containing a weight upto 500 kg?

(A) Atleast 5 individual samples
(B) One individual sample for each 300 kg, but not less than 5 samples
(C) One individual sample for each 500 kg, but not less than 10 samples
(D) One individual sample for each 700 kg, but not less than 40 samples

34. If the seed lot having a weight from 501 to 3000 kg, how many primary samples are to be taken?

(A) Atleast 5 individual samples
(B) One individual sample for each 300 kg, but not less than 5 samples.
(C) One individual sample for each 500 kg, but not less than 10 samples.
(D) One individual sample for each 700 kg, but not less than 40 samples.

35. What is the size of submitted sample for sesamum

(A) 60 g (B) 70 g
(C) 40 g (D) 50 g

36. Which one of the following weight is the size of submitted sample for fingermillet

(A) 60 g (B) 70 g
(C) 40 g (D) 50 g

37. ---------------- is the size of submitted sample for lucerne

(A) 60 g (B) 70 g
(C) 40 g (D) 50 g

38. What is the size of submitted sample for Chinese cabbage

(A) 60 g (B) 70 g
(C) 40 g (D) 50 g

39. Which one of the following is the submitted sample size of groundnut pods

(A) 250 g (B) 500 g
(C) 750 g (D) 1000 g

40. What is the size of submitted sample for bittergourd

(A) 1000 g (B) 750 g
(C) 500 g (D) 250 g

41. -------------- is the submitted sample size of tomato varieties

(A) 700 g (B) 350 g
(C) 150 g (D) 70 g

42. What is the size of submitted sample for pumpkin

(A) 700 g (B) 350 g
(C) 150 g (D) 70 g

43. Which one of the following is the size of submitted sample for brinjal

(A) 700 g (B) 350 g
(C) 150 g (D) 70 g

44. What is the size of submitted sample of ashgourd

(A) 700 g (B) 350 g
(C) 150 g (D) 70 g

45. What is the size of submitted sample for tomato hybrids

(A) 7 g (B) 30 g
(C) 80 g (D) 100 g

46. ------------- is the size of submitted sample of cauliflower

(A) 7 g (B) 30 g
(C) 80 g (D) 100 g

47. Which one of the following is the size of submitted sample of onion

(A) 7 g (B) 30 g
(C) 80 g (D) 100 g

48. ----------- is the size of submitted sample of carrot

(A) 7 g (B) 30 g
(C) 80 g (D) 100 g

49. What is the size of working sample for purity analysis for okra

(A) 140 g (B) 70 g
(C) 120 g (D) 60 g

50. --------- is the size of working sample for purity analysis of sunnhemp

(A) 140 g (B) 70 g
(C) 120 g (D) 60 g

ANSWER SHEET

1. **A**	2. **B**	3. **D**	4. **C**	5. **C**	6. **A**	7. **D**	8. **B**	9. **A**	10. **D**
11. **B**	12. **D**	13. **C**	14. **A**	15. **D**	16. **B**	17. **C**	18. **C**	19. **D**	20. **B**
21. **A**	22. **C**	23. **C**	24. **D**	25. **A**	26. **B**	27. **A**	28. **D**	29. **B**	30. **C**
31. **C**	32. **D**	33. **A**	34. **B**	35. **B**	36. **A**	37. **D**	38. **C**	39. **D**	40. **A**
41. **D**	42. **B**	43. **C**	44. **A**	45. **A**	46. **D**	47. **C**	48. **B**	49. **A**	50. **B**

Chapter 7

MIXING AND DIVIDING OF SEED

1. Which one of the following divider is working on the principle of gravitational force?

   (A) Seed divider (B) Gamet divider

   (C) Boerner divider (D) Spoon divider

2. The principle involved in the dividing and mixing of seed sample using ------------ is centrifugal force.

   (A) Seed divider (B) Gamet divider

   (C) Boerner divider (D) Spoon divider

3. Which one of the principle is followed for mixing and dividing of seeds using Boerner divider?

   (A) Centrifugal force

   (B) Electrical force

   (C) Magnetic force

   (D) Gravitational force

4. Which principle is followed for mixing and dividing of seeds using gamet divider?

   (A) Centrifugal force

   (B) Electrical force

   (C) Magnetic force

   (D) Gravitational force

5. Which one of the following method is used for mixing and dividing of seeds requiring working sample upto 10 g.

   (A) Spoon method

   (B) Random cup method

   (C) Modified halving method

   (D) Hand halving method

6. Which one of the following method is used for mixing and dividing of seeds of small seeded species?

   (A) Spoon method

   (B) Random cup method

   (C) Modified halving method

   (D) Hand halving method

7. Which one of the following method is used for mixing and dividing of chaffy seeds?

   (A) Spoon method

   (B) Random cup method

   (C) Modified halving method

   (D) Hand halving method

8. Which one of the following method is suitable for seeds that are not extremely chaffy and do not bounce or roll?

   (A) Spoon method

   (B) Random cup method

   (C) Modified halving method

   (D) Hand halving method

ANSWER SHEET

1. **C** 2. **B** 3. **D** 4. **A** 5. **B** 6. **A** 7. **D** 8. **B**

Chapter 8

PURITY ANALYSIS

1. Seed purity analysis is done on ------------
   (A) Submitted sample
   (B) Primary sample
   (C) Composite sample
   (D) Working sample
2. Seed testing is carried out in STL in accordance with ------------------ rules.
   (A) ISTA (B) AOSA
   (C) ISST (D) ICAR
3. Which one of the following is not relevant to pure seed fraction?
   (A) All botanical varieties of that kind / species
   (B) Broken seeds, if the size is > ¾ of the original size
   (C) Broken seeds, if the size is > ½ of the original size
   (D) Immature, shrivelled seeds
4. What is the number of places of decimal upto which each of the components of purity analysis must be weighed and reported, if the working sample weight is > 100 g to < 1000 g?
   (A) 1 (B) 2
   (C) 3 (D) 4
5. The number of places of decimals is -------, if the working sample weight is > 10 g to < 100 g while reporting the different components of seed purity analysis.
   (A) 1 (B) 2
   (C) 3 (D) 4
6. Which one of the decimal place is followed, if the purity analysis is done is working seed sample weighing > 1 g to < 10 g while reporting the results?
   (A) 1 (B) 2
   (C) 3 (D) 4
7. If the working sample size is > 1g, how many decimal places are followed while reporting the results of purity analysis
   (A) 1 (B) 2
   (C) 3 (D) 4
8. Freedom from inert matter and defective seeds :
   (A) Genetic purity (B) Physical purity
   (C) Defective purity (D) Normal purity
9. Seed test that provides information on weed seed and inert material is called :
   (A) Clean seed test
   (B) Purity test
   (C) Wholesomeness test
   (D) Viability test
10. Which of the following is not a valid classification of weeds in a seed purity test?
    (A) Common weed (B) Noxious weed
    (C) Annual weed (D) Restricted

ANSWER SHEET

1. D	2. A	3. B	4. A	5. B	6. C	7. D	8. D	9. B	10. C

Chapter 9

SEED MOISTURE DETERMINATION

1. On what basis, the moisture content of seed sample is expressed in percentage?
   (A) Dry basis
   (B) Wet basis
   (C) Water basis
   (D) Moisture basis

2. Moisture content of the seed should be determined within ---------------- minutes after opening of container.
   (A) 1 (B) 2
   (C) 3 (D) 4

3. Based on which, the moisture content of seeds is determined
   (A) Electrical conductivity
   (B) Moisture conductivity
   (C) Physical conductivity
   (D) Mechanical conductivity

4. Which one of the following is the indirect method of seed moisture determination?
   (A) Toluene distillation method
   (B) Air oven method
   (C) Moisture meter method
   (D) Lyophilization method

5. Which one of the following method is not the direct method of seed moisture estimation?
   (A) Toluene distillation method
   (B) Air oven method
   (C) Moisture meter method
   (D) Lyophilization method

6. Which one of the following seed requires fine grinding for seed moisture estimation?
   (A) Groundnut
   (B) Soybean
   (C) Field bean
   (D) Pearl millet

7. Which one of the following seed requires coarse grinding before drying for moisture determination?
   (A) Pearl millet
   (B) Sorghum
   (B) Soybean
   (D) Finger millet

8. If the seed size is less than 8mm, what will be the weight of seed sample taken for seed moisture determination?
   (A) 3-4 g (B) 4-5 g
   (C) 9 g (D) 10 g

9. What will be the seed sample weight for moisture determination, if the seed size is more than 8 mm?
   (A) 3-4 g (B) 4-5 g
   (C) 9 g (D) 10 g

10. What is the temperature and duration of oven drying for moisture estimation using low constant temperature method?
    (A) 101 ± 2°C for 17 ± 1h
    (B) 102 ± 2°C for 17 ± 1h
    (C) 104 ± 2°C for 17 ± 1h
    (D) 103 ± 2°C for 17 ± 1h

11. At what temperature and duration, the seeds are dried in oven using high constant temperature method for moisture determination?

(A) 110° ± 3°C for 4h

(B) 110° ± 3°C for 6h

(C) 130° ± 2°C for 4h

(D) 130° ± 3°C for 4h

12. The seed sample for moisture estimation should be weighed in ---------------- decimals.

(A) 0 (B) 1

(C) 2 (D) 3

13. What method is adopted for moisture determination of seeds which contain volatile oils?

(A) Moisture meter method

(B) Air oven method

(C) Toluene distillation method

(D) Lyophilization method

ANSWER SHEET

1. **B**	2. **B**	3. **A**	4. **C**	5. **C**	6. **D**	7. **C**	8. **B**	9. **D**	10. **D**
11. **D**	12. **D**	13. **C**							

Chapter 10

GERMINATION TEST

1. Which one of the following method is adopted for germination test of light requiring seeds?

   (A) Between paper
   (B) Top of paper
   (C) Sand medium
   (D) Soil medium

2. As per ISTA rules, how many seeds are used for conducting germination test?

   (A) 100 (B) 200
   (C) 300 (D) 400

3. Which method is used for germination test of large seeds?

   (A) Between paper
   (B) Top of paper
   (C) Sand medium
   (D) Soil medium

4. Which one of the following is not relevant for seedling evaluation?

   (A) Normal seedlings
   (B) Abnormal seedlings
   (C) Hard seeds
   (D) Fresh seeds

5. Which one of the following is not pertaining to the normal seedling characteristics?

   (A) A well developed root system
   (B) A well developed shoot system
   (C) One cotyledon in monocots and one cotyledon in dicots
   (D) A well developed epicotyl

6. Which one of the following is not relevant to abnormal seedling?

   (A) Decayed seedling, but the parent seed is not the source of infection
   (B) Decayed seedling
   (C) Damaged seedling
   (D) Deformed seedling

7. Which category of seed yields the milky substance when it is pressed at the end of germination test?

   (A) Hard seed
   (B) Imbibed seed
   (C) Dead seed
   (D) Fresh ungerminated seed

8. Which category of seed does not imbibe water when evaluated at the end of germination test?

   (A) Hard seed
   (B) Imbibed seed
   (C) Dead seed
   (D) Fresh ungerminated seed

9. ------------------ needs dormancy breaking treatment for germination, as observed during germination test

   (A) Fresh seed
   (B) Imbibed seed
   (C) Dead seed
   (D) Fresh ungerminated seed

10. Pure Live Seed (PLS) is related to :

(A) Physical purity

(B) Genetic purity

(C) Germination percentage

(D) Contamination

11. The number of healthy normal seedlings / number of seed tested × 100

(A) Percent viability

(B) Percent germination

(C) Percent purity

(D) Percent healthy seed

12. Percent live seed = (Y × % purity) / 100. What is "Y"?

(A) % crop seed

(B) % germination

(C) % viability

(D) % weed

ANSWER SHEET

1. **B**	2. **D**	3. **C**	4. **D**	5. **C**	6. **A**	7. **C**	8. **A**	9. **D**	10. **C**
11. **B**	12. **B**								

Chapter 11

QUICK VIABILITY TEST

1. Which one of the following chemical is used for assessing the viability of seeds?
(A) 2,3,2 Triphenyl Tetrazolium Chloride
(B) 2,3,3 Triphenyl Tetrazolium Chloride
(C) 2,3,4 Triphenyl Tetrazolium Chloride
(D) 2,3,5 Triphenyl Tetrazolium Chloride

2. Which chemical process is taking place in the living cells and accepts hydrogen from the hydrogenases during TZ test?
(A) Reduction process
(B) Oxidation process
(C) Peroxidation process
(D) Hydrogenation process

3. Which colour is produced for evaluation of viable and non viable seeds during TZ test?
(A) Light red (B) Red
(C) Dark red (D) Very dark red

4. 2,3,5 - Triphenyl Tetrazolium Chloride solution is ------------------- in colour.
(A) Colourless (B) White
(C) Red (D) Pink

5. Which one of the following is correct for preparation of 1% TZ solution?
(A) 10 g TZ in 100 ml of water
(B) 1 g TZ in 10000 ml of water
(C) 1 g TZ in 100 ml of water
(D) 1 g TZ in 1000 ml of water

6. Which one of the following is optimum pH of TZ solution for proper staining of living part of seed
(A) 4 (B) < 4
(C) 7 (D) > 7

7. Which of the following is correct for staining of dead parts of seed?
(A) Red colour dead ones
(B) Pink colour dead ones
(C) Colourless dead ones
(D) White colour dead ones

8. In one litre buffer solution, how many grams of TZ salt should be mixed to prepare 1% solution.
(A) 0.1 g (B) 1 g
(C) 10 g (D) 100 g

9. For preconditioning of seed for TZ test, slow moistening is given for ---------- seeds.
(A) Sorghum (B) Soybean
(C) Maize (D) Paddy

10. Which one of the following method is suitable for preconditioning of maize seed for TZ test?
(A) Bisect longitudinally
(B) Bisect laterally
(C) Remove seed coat
(D) No conditioning

11. For guinea grass, the preconditioning method used for TZ test is ------------------.

(A) Bisect longitudinally

(B) Bisect laterally

(C) Remove seed coat

(D) No conditioning

12. ---------------------- is the preconditioning method adopted for castor for TZ test.

(A) Bisect longitudinally

(B) Bisect laterally

(C) Remove seed coat

(D) No conditioning

13. Which one of the following method is suggested for preconditioning of horsegram seed for TZ test?

(A) Bisect longitudinally

(B) Bisect laterally

(C) Remove seed coat

(D) No conditioning

14. Which one of the following is not correct for germinable seed while evaluating TZ test.

(A) Plumule unstained

(B) Entire embryo stained bright red

(C) Extremeties of scutellum unstained

(D) Extremeties of radicle tip unstained

15. ---------- is the non-germinable seed while evaluating the TZ test.

(A) Plumule unstained

(B) Entire embryo stained bright red

(C) Extremeties of scutellum unstained

(D) Extremeties of radicle tip unstained

ANSWER SHEET

1. **D**	2. **A**	3. **B**	4. **A**	5. **C**	6. **C**	7. **C**	8. **C**	9. **B**	10. **A**
11. **B**	12. **C**	13. **D**	14. **A**	15. **A**					

Chapter 12

GROW OUT TEST

1. What test is being conducted to evaluate the genetic purity of seed?

   (A) Vigour test (B) Grow out test
   (C) Variety test (D) DUS test

2. Grow out test is must be conducted for ---------------------.

   (A) Greengram (B) Finger millet
   (C) Cotton (D) Horsegram

3. What is the seed sample size for grow out test?

   (A) 200 seeds (B) 400 seeds
   (C) 600 seeds (D) 800 seeds

4. What is the maximum permissible limit of offtypes for 99.9% genetic purity (minimum)?

   (A) 0.10% (B) 0.20%
   (C) 0.30% (D) 0.50%

5. -------------- is the maximum permissible off types for 99.8% genetic purity.

   (A) 0.10% (B) 0.20%
   (C) 0.30% (D) 0.50%

6. Which one of the following is the maximum permissible off types for 99.7% genetic purity?

   (A) 0.10% (B) 0.20%
   (C) 0.30% (D) 0.50%

7. What is the maximum permissible off types permitted for 99.5% genetic purity?

   (A) 0.10% (B) 0.20%
   (C) 0.30% (D) 0.50%

ANSWER SHEET

1. **B** 2. **C** 3. **B** 4. **A** 5. **B** 6. **C** 7. **D**

Chapter 13

SEED VIGOUR

1. Which one of the following is the physical vigour test?
   (A) Cold test
   (B) TZ test
   (C) Seed size
   (D) Vigour index
2. ------------ is the biochemical test for evaluating seed vigour.
   (A) Cold test
   (B) TZ test
   (C) Seed size
   (D) Vigour index
3. Which is the performance test for vigour evaluation?
   (A) Cold test
   (B) TZ test
   (C) Seed size
   (D) Vigour index
4. Which one of the following is the stress test for seed vigour evaluation?
   (A) Cold test
   (B) TZ test
   (C) Seed size
   (D) Vigour index
5. Measurement of seedling length is the -------------- test for evaluating seed vigour.
   (A) Physical test
   (B) Performance test
   (C) Biochemical test
   (D) Stress test
6. Which type of vigour test is GADA test?
   (A) Physical test
   (B) Performance test
   (C) Biochemical test
   (D) Stress test
7. Brick gravel test is
   (A) Physical test
   (B) Performance test
   (C) Biochemical test
   (D) Stress test
8. Measurement of seed density is used to evaluate ---------------- of seed vigour.
   (A) Physical test
   (B) Performance test
   (C) Biochemical test
   (D) Stress test
9. Who has worked out the vigour index?
   (A) Hiltner
   (B) Moore
   (C) Abdul-Baki and Anderson
   (D) Mathews and Bradnock
10. Which one of the following scientist found that electrical conductivity is positively correlated with field emergence?
   (A) Hiltner
   (B) Moore
   (C) Abdul-Baki and Anderson
   (D) Mathews and Bradnock

11. Who has contemplated the TZ test for evaluation of vigour of a seed lot?

(A) Hiltner

(B) Moore

(C) Abdul-Baki and Anderson

(D) Mathews and Bradnock

12. Who is the scientist responsible for formulating the procedure for brick gravel test?

(A) Hiltner

(B) Moore

(C) Abdul-Baki and Anderson

(D) Mathews and Bradnock

13. Cold test is used to evaluate seed vigour of ----------------- seed.

(A) Maize

(B) Sorghum

(C) Paddy

(D) Wheat

14. For conducting osmotic stress, which chemical is used

(A) Potassium chloride

(B) Sodium chloride

(C) Inositol

(D) Mannitol

15. What is the suitable temperature and relative humidity for accelerated ageing test?

(A) 40±2°C and 95±2% RH

(B) 40±2°C and 90±2% RH

(C) 41±2°C and 95±2% RH

(D) 41±2°C and 90±2% RH

ANSWER SHEET

1. **C**	2. **B**	3. **D**	4. **A**	5. **B**	6. **C**	7. **D**	8. **A**	9. **C**	10. **D**
11. **B**	13. **A**	13. **A**	14. **D**	15. **B**					

Chapter 14

VARIETY IDENTIFICATION

1. Which one of the following is not relevant to varietal identification?
   (A) To evaluate genetic purity
   (B) To implement protection of plant breeders and farmers right
   (C) To evaluate the physical purity
   (D) To conserve germplasm and genetic diveristy
2. There are few approaches to varietal identification which are furnished below. Find out the incorrect one.
   (A) Pathological approaches
   (B) Physiological approaches
   (C) Biochemical approaches
   (D) Molecular approaches
3. Which one of the following seed has beak that is used for variety identification?
   (A) Redgram (B) Castor
   (C) Paddy (D) Groundnut
4. The best example for awned seed is --------
   (A) Redgram (B) Castor
   (C) Paddy (D) Groundnut
5. Which one of the following kind of seed is identified based on the presence of caruncle
   (A) Redgram (B) Castor
   (C) Paddy (D) Groundnut
6. Which one of the following seed has hilum?
   (A) Redgram (B) Castor
   (C) Paddy (D) Groundnut
7. Which one of the following crop seed responds well for phenol test?
   (A) Sorghum (B) Wheat
   (C) Pearl millet (D) Maize
8. In sodium hydroxide test, white and red varieties of rice becomes ----------- colour.
   (A) Bright yellow (B) Light yellow
   (C) Dark red (D) Dull red
9. Sodium hydroxide test is used to identify red and white varieties of ------------------
   (A) Rice (B) Sorghum
   (C) Maize (D) Pearl millet
10. The presence or absence of tannic acid in sorghum seed is determined by --------test.
    (A) Sodium hydroxide
    (B) Potassium dichromate
    (C) Phenol
    (D) KOH bleach
11. Ferrous sulphate colour test is recommended for ------------- seeds.
    (A) Wheat (B) Paddy
    (C) Sorghum (D) Maize

ANSWER SHEET

1. C 2. A 3. D 4. C 5. B 6. A 7. B 8. A 9. A 10. D 11. C

Chapter 15

INTERNATIONAL SEED TESTING ASSOCIATION (ISTA)

1. Which year, the ISTA was established?
   (A) 1904 (B) 1914
   (C) 1924 (D) 1934
2. In which country, the ISTA is located?
   (A) Germany (B) Switzerland
   (C) India (D) USA
3. Zurich is the headquarters for ---------------
   (A) AOSA (B) ISST
   (C) ISTA (D) UPOV
4. What colour tag is used for International seed lot certificate?
   (A) White (B) Green
   (C) Blue (D) Orange
5. What colour tag is used for International seed sample certificate?
   (A) White (B) Green
   (C) Blue (D) Orange
6. ISTA publishes -----------------
   (A) Journal Seed Technology
   (B) Seed Science and Technology
   (C) Seed Research
   (D) Journal of Seed Technology
7. International Seed Testing Association was organized on :
   (A) 10-Jul-24 (B) 12-Jul-25
   (C) 12-Feb-42 (D) 10-Feb-24

ANSWER SHEET

1. **C** 2. **B** 3. **C** 4. **D** 5. **C** 6. **B** 7. **A**

Unit 5
SEED STORAGE

Chapter 1

SEED STORAGE

1. Which one of the following is not relevant to principles of seed storage?
   (A) Storage of poor quality seed
   (B) Storage environment should be cool and dry
   (C) Proper seed store sanitation
   (D) Effective control of storage pests
2. For storage of recalcitrant seeds, at what moisture content they are maintained
   (A) 20-25% (B) 25-30%
   (C) 30-35% (D) 35-40%
3. Which one of the following seed is a recalcitrant seed?
   (A) Guava (B) Pomegranate
   (C) Grapes (D) Mango
4. Which one of the following seed is desiccation intolerant?
   (A) Rubber (B) Pepper
   (C) Coriander (D) Black pepper
5. The best example for desiccation tolerant seed is --------------.
   (A) Rubber (B) Citrus
   (C) Maize (D) Mango
6. The short lived seeds are known as ---------
   (A) Megabiotic seed
   (B) Macrobiotic seed
   (C) Mesobiotic seed
   (D) Microbiotic seed
7. ------------------- is medium lived seed.
   (A) Megabiotic seed
   (B) Macrobiotic seed
   (C) Mesobiotic seed
   (D) Microbiotic seed
8. The long lived seed is called as -----------
   (A) Megabiotic seed
   (B) Macrobiotic seed
   (C) Mesobiotic seed
   (D) Microbiotic seed
9. Which one of the following pest is a pre harvest infester?
   (A) Wasps (B) Beetles
   (C) Bruchids (D) Moths
10. Which one of the following statement is correct with respect to seed moisture?
    (A) For every 1% decrease in seed moisture, the storage life of seed doubled, provided the seed moisture content is 5-14%
    (B) For every 1% decrease in seed moisture, the storage life of seed doubled, provided the seed moisture content is 4-14%
    (C) For every 1% decrease in seed moisture, the storage life of seed doubled, provided the seed moisture content is 5-15%
    (D) For every 1% decrease in seed moisture, the storage life of seed doubled, provided the seed moisture content is 6-14%

11. For every decrease of ------------------ in storage temperature, the seed storage life doubled.

(A) 5°F (B) 10°F
(C) 15°F (D) 20°F

12. A good seed storage is achieved when the RH in percentage and temperature in Fahrenheit of storage environment adds upto

(A) 110 (B) 90
(C) 100 (D) 80

13. Which one of the following is a moisture vapour proof container?

(A) Paper bags
(B) Cloth bags
(C) Polyvinyl bags
(D) > 700 gauge polyethylene bag

14. ------------------ is a moisture impervious, but vapour pervious container

(A) Cloth bags
(B) Gunny bags
(C) < 100 gauge polyethylene bag
(D) > 700 gauge polyethylene bag

15. What is the optimum moisture content for fumigation?

(A) > 14% (B) < 14%
(C) > 15% (D) < 15%

16. At what height, the seed bags are stacked in storage godown?

(A) 6-8 tires (B) 7-9 tires
(C) 5-7 tires (D) 4-6 tires

17. What pesticides should be sprayed to disinfect the storage godown before storage?

(A) Parathion (B) Malathion
(C) Rogar (D) Endosulfan

18. Germination of 101 years old seeds was observed in

(A) Melon family
(B) Arctic Tundra Lupine
(C) Moth Mullein
(D) Hesperidiums

19. Seed moisture for storage varies from crop to crop in ranges from

(A) 15-20% (B) 30-40%
(C) 1-2% (D) 9-12%

20. In cryogenic storage, seeds are stored at

(A) 75°C (B) -196°C
(C) -100°C (D) -65°C

ANSWER SHEET

1. **A**	2. **C**	3. **D**	4. **A**	5. **C**	6. **D**	7. **C**	8. **B**	9. **C**	10. **A**
11. **B**	12. **C**	13. **D**	14. **C**	15. **B**	16. **A**	17. **B**	18. **C**	19. **D**	20. **B**

Chapter 2

SEED DETERIORATION

1. Which one of the following is not related to seed deterioration?

   (A) It is a deteriorative changes
   (B) It is an inexorable process
   (C) It is a irreversible process
   (D) It is a reversible process

2. Find out the incorrect statement with respect to changes associated with seed deterioration?

   (A) Mutational changes
   (B) Genetic changes
   (C) Biochemical changes
   (D) Physiological changes

3. Which one of the following is a ultrastructural change associated with seed deterioration?

   (A) Reduced respiration
   (B) Mitochondrial fragmentation
   (C) Seed colour changes
   (D) Delayed germination

4. ---------------- is the physiological changes associated with seed deterioration?

   (A) Reduced respiration
   (B) Mitochondrial fragmentation
   (C) Seed colour changes
   (D) Delayed germination

5. Which one of the following is a morphological change associated with seed deterioration?

   (A) Reduced respiration
   (B) Mitochondrial fragmentation
   (C) Seed colour changes
   (D) Delayed germination

6. Which one of the following is a biochemical change associated with seed deterioration?

   (A) Reduced respiration
   (B) Mitochondrial fragmentation
   (C) Seed colour changes
   (D) Delayed germination

7. Which one of the following is the old consensus of seed deterioration?

   (A) Impaired transcription leading to incomplete enzyme synthesis
   (B) Production of free radicals causing damages to the cell
   (C) Hindering of mitochondrial replication
   (D) Lesser ATP synthesis

8. Which one of the following is not related to new consensus of seed deterioration?

   (A) Impaired transcription leading to incomplete enzyme synthesis
   (B) Production of free radicals causing damages to the cell
   (C) Hindering of mitochondrial replication
   (D) Lesser ATP synthesis

9. Which one of the following is the new consensus of seed deterioration?
   (A) Impaired transcription leading to incomplete enzyme synthesis
   (B) Production of free radicals causing damages to the cell
   (C) Degraded DNA fail to produce mRNA
   (D) Degrading of stored mRNA

ANSWER SHEET

1. **D** 2. **A** 3. **B** 4. **D** 5. **C** 6. **A** 7. **A** 8. **B** 9. **C**

Chapter 3

SEED TREATMENT

1. Seed treatment with bavistin given before sowing is to control ------------------.
   (A) Seed borne virus
   (B) Seed borne bacteria
   (C) Soil borne virus
   (D) Seed borne fungal pathogens

2. What is the dosage of carbendazim for seed treatment?
   (A) 1-3 g/kg of seeds
   (B) 2-4 g/kg of seeds
   (C) 3-5 g/kg of seeds
   (D) 4-6 g/kg of seeds

3. To control bruchids, how much quantity of activated clay is treated?
   (A) 25 g/kg of seed
   (B) 50 g/kg of seeds
   (C) 75 g/kg of seed
   (D) 100 g/kg of seeds

4. For slurry treatment, how much quantity of water is added to the pesticides?
   (A) 5 ml (B) 10 ml
   (C) 15 ml (D) 20 ml

5. Which one of the following treatment is not a storage treatment?
   (A) Halogenation
   (B) Antioxidant treatment
   (C) Fumigation
   (D) Magnetic treatment

6. For halogenation treatment, -------------- is a good carrier.
   (A) Calcium chloride
   (B) Calcium carbonate
   (C) Calcium bicarbonate
   (D) Calcium oxychloride

7. Which one of the following is a bleaching powder?
   (A) Calcium chloride
   (B) Calcium carbonate
   (C) Calcium bicarbonate
   (D) Calcium oxychloride

8. Which one of the following is an antioxidant?
   (A) Vitamin E (B) Vitamin B
   (C) Vitamin D (D) Vitamin F

9. Antioxidant namely vit. E provides ------------------- atom for quenching free radical
   (A) Oxygen (B) Carbon
   (C) Hydrogen (D) Chlorine

10. Eradication of fungal pathogen which are deep-seated inside the seed tissues is known as-----------
    (A) Seed protection
    (B) Seed disinfection
    (C) Seed disinfestation
    (D) Seed fumigation

11. Control of fungal pathogen infected on the surface of seed coat is named as ------------

(A) Seed protection
(B) Seed disinfection
(C) Seed disinfestation
(D) Seed fumigation

12. Exposing of gaseous form of chemicals to control storage pests is called as -------------

(A) Seed protection
(B) Seed disinfection
(C) Seed disinfestation
(D) Seed fumigation

13. What is the optimum moisture content of seed for undertaking fumigation?

(A) Above 16% (B) Below 14%
(C) Above 14% (D) More than 14%

14. The optimum dosage of aluminium phosphide used for fumigation.

(A) 1-2 g cu.m (B) 2-3 g cu.m
(C) 3-4 g cu.m (D) 4-5 g cu.m

15. Which one of the following is a fumigant?

(A) Calcium cyanide
(B) Calcium chloride
(C) Calcium carbonate
(D) Calcium oxychloride

16. Which one of the following is not related to hydration and dehydration treatment?

(A) Given only to stored seeds
(B) Given to low and medium vigour non-leguminous seeds
(C) Given to fresh seeds
(D) Avoid direct soaking of leguminous seed

17. Which one of the following is not a chemical used for soaking and drying?

(A) Potassium phosphate
(B) Sodium dihydrogen phosphate
(C) Potassium hydroxide
(D) Sodium chloride

18. In hydration and dehydration treatment, the seeds are soaked to raise the seed moisture upto -----------------

(A) 15-20% (B) 20-25%
(C) 25-30% (D) 30-35%

19. In moisture equilibration treatment, the seeds are exposed to -------------- RH.

(A) 110% (B) 100%
(C) 90% (D) 80%

20. What is the moisture content of sand in which the seeds are conditioned and dried?

(A) 0-5% (B) 5-10%
(C) 10-15% (D) 15-20%

ANSWER SHEET

1. **D**	2. **B**	3. **D**	4. **A**	5. **D**	6. **B**	7. **D**	8. **A**	9. **C**	10. **B**
11. **C**	13. **D**	13. **B**	14. **B**	15. **A**	16. **C**	17. **C**	18. **C**	19. **B**	20. **B**

Chapter 4

GERMPLASM CONSERVATION AND CRYOPRESERVATION

1. Which one of the following is *in situ* conservation?
   (A) Gene banks
   (B) Field banks
   (C) *in vitro* plant tissue collection
   (D) Conserving in natural habitat

2. Which one of the following is not related to *ex situ* conservation?
   (A) Gene banks
   (B) Field banks
   (C) *in vitro* plant tissue collection
   (D) Conserving in natural habitat

3. What is the optimum moisture content of seed that is to be conserved in seed bank?
   (A) 5% (B) 4%
   (C) 7% (D) 6%

4. In seed bank, the seeds are preserved at -------------- temperature.
   (A) 18°C (B) -18°C
   (C) 19°C (D) -19°C

5. Where Royal Botanic Garden is located?
   (A) USA (B) Germany
   (C) London (D) Paris

6. Which country is maintaining the Svalbard Global seed valut?
   (A) USA (B) Norway
   (C) London (D) Germany

7. Where IRRI is located?
   (A) Mexico (B) Columbia
   (C) Phillippines (D) Peru

8. In which country, International Centre for Potato is located?
   (A) Mexico (B) Columbia
   (C) Phillippines (D) Peru

9. International Centre for Tropical Agriculture is located at ------------.
   (A) Mexico (B) Columbia
   (C) Phillippines (D) Peru

10. Which one of the following country is the headquarters for CIMMYT?
    (A) Mexico (B) Columbia
    (C) Phillippines (D) Peru

11. Which organisation is responsible for paddy germplasm collection and maintenance?
    (A) IRRI (B) CIMMYT
    (C) IITA (D) CIP

12. The cassava germplasm is conserved in
    (A) IRRI (B) CIMMYT
    (C) IITA (D) CIP

13. Where potato germplasm is conserved?
    (A) IRRI (B) CIMMYT
    (C) IITA (D) CIP

14. Which one of the following organisation is conserving the wheat germplasm?

(A) IRRI (B) CIMMYT
(C) IITA (D) CIP

15. The chemical used for cryopreservation is - --------------.

(A) Liquid hydrogen
(B) Liquid oxygen
(C) Liquid nitrogen
(D) Liquid carbon

16. At which temperature, the plant cells and tissues are cryopreserved?

(A) Low temperature
(B) Ultra low temperature
(C) High temperature
(D) Ultra high temperature

17. In cryopreservation, the liquid nitrogen is maintained at -------------- temperature to conserve the cells, tissues and organs

(A) -196°C (B) -190°C
(C) -186°C (D) -180°C

18. Which one of the following is not stored under cryopreservation?

(A) Anther and pollen
(B) Ovary and ovule
(C) Shoot apex
(D) Meristem tip

19. Which one of the following is not a cryoprotectant?

(A) Sugars (B) PEG
(C) Alcohols (D) Mannitol

20. DMSO (Dimethyl sulfoxide) is used as

(A) Gelling agent
(B) Alkaylating agent
(C) Chelating agent
(D) Cryoprotectant

21. Germplasm means

(A) A breeding method
(B) A collection of genetic material
(C) A new disease
(D) A new clone

22. Cryopreservation is not suitable for preserving organisms which are :

(A) Aerobic (B) Anaerobic
(C) Frost resistant (D) Frost sensitive

23. National Bureau of Plant Genetic Resource (NBPGR) was established in:

(A) 1960 (B) 1971
(C) 1976 (D) 1980

24. International Rice Research Institute (IRRI) is located in:

(A) Mexico (B) Chaina
(C) Philippines (D) India

ANSWER SHEETS

1. **D**	2. **D**	3. **A**	4. **B**	5. **C**	6. **B**	7. **C**	8. **D**	9. **B**	10. **A**
11. **A**	12. **C**	13. **D**	14. **B**	15. **C**	16. **B**	17. **A**	18. **B**	19. **D**	20. **D**
21. **B**	22. **A**	23. **C**	24. **C**						

Unit 6
SEED HEALTH

Chapter 1

SEED PATHOLOGY

1. Which one of the following is not related to pathogens that are not causing diseases?
(A) Insects (B) Fungi
(C) Bacteria (D) Virus

2. Which one of the following pathogens infects the developing embryo of seeds by travelling through the veins of the plant and then through the vascular system?
(A) Insects (B) Fungi
(C) Bacteria (D) Virus

3. Fungal pathogens that persist on the plant residues are known as -----------------
(A) Seed borne pathogen
(B) Air borne pathogen
(C) Water borne pathogen
(D) Soil borne pathogen

4. Which one of the following pathogen causes root or stem disruption that blocks the uptake of water and nutrients?
(A) Seed borne pathogen
(B) Air borne pathogen
(C) Water borne pathogen
(D) Soil borne pathogen

5. ------------- causes wilting of plant.
(A) Seed borne pathogen
(B) Air borne pathogen
(C) Water borne pathogen
(D) Soil borne pathogen

6. Which one of the following pathogen produces oospores that persist for long period?
(A) Seed borne pathogen
(B) Air borne pathogen
(C) Water borne pathogen
(D) Soil borne pathogen

7. Root rot is caused by ---------------------
(A) Cylindrocladium
(B) Verticillium
(C) Pythium (D) Aspergillus

8. The casual pathogen of wilt is --------------
(A) Cylindrocladium
(B) Verticillium
(C) Pythium (D) Aspergillus

9. Which one of the following fungi causes damping off disease?
(A) Cylindrocladium
(B) Verticillium
(C) Pythium
(D) Aspergillus

10. Which fungi infects the seed through seed parts?
(A) *Fusarium oxysporum*
(B) *Xanthomonas compestris*
(C) *Ustilago tritici*
(D) *Rhizoctonia solani*

11. Which pathogen enters the seed through the wall of ovary?
(A) *Fusarium oxysporum*
(B) *Xanthomonas compestris*
(C) *Ustilago tritici*
(D) *Rhizoctonia solani*

12. ---------------- penetrates through the stigma and infects the seed internally.

(A) *Fusarium oxysporum*
(B) *Xanthomonas compestris*
(C) *Ustilago tritici*
(D) *Rhizoctonia solani*

13. Which one of the following fungi is responsible for systemic infection through fruits?

(A) *Fusarium oxysporum*
(B) *Xanthomonas compestris*
(C) *Ustilago tritici*
(D) *Rhizoctonia solani*

14. Which one of the following fungi penetrates through the wounds and infects the seed internally?

(A) *Ustilago tritici*
(B) *Pseudomonas syringae*
(C) *Ascochyta pinodella*
(D) *Ustilago nuda*

15. --------------- is the pathogen that stick to the surface of the seed?

(A) *Ustilago tritici*
(B) *Pseudomonas syringae*
(C) *Ascochyta pinodella*
(D) *Ustilago nuda*

16. Which one of the following causes systemic infection of the embryo?

(A) Seed borne virus
(B) Water borne fungi
(C) Soil borne fungi
(D) Air borne fungi

17. --------------- is the fungi that causes non embryo infection, but systemic

(A) Sclerospora
(B) *Colletotrichum truncatum*
(C) *Neovossia indica*
(D) *Corynebacterium michiganense*

18. Which pathogen contaminates episperm and that is systemic transmission?

(A) Sclerospora
(B) *Colletotrichum truncatum*
(C) *Neovossia indica*
(D) *Corynebacterium michiganense*

19. In soybean, ------------- infects the embryo, but it is non systemic transmission.

(A) Sclerospora
(B) *Colletotrichum truncatum*
(C) *Neovossia indica*
(D) *Corynebacterium michiganense*

20. Which one of the following fungi is responsible for non systemic infection of episperm?

(A) Sclerospora
(B) *Colletotrichum truncatum*
(C) *Neovossia indica*
(D) *Corynebacterium michiganense*

21. What is the designated disease of paddy?

(A) Grain smut (B) Head smut
(C) Loose smut (D) Bunt

22. The designated disease of pearl millet is -----------

(A) Grain smut (B) Head smut
(C) Loose smut (D) Bunt

23. Which one of the following is the designated disease of sorghum?

(A) Grain smut (B) Head smut
(C) Loose smut (D) Bunt

24. ------------------------ is the designated disease of wheat?

(A) Grain smut (B) Head smut
(C) Loose smut (D) Bunt

25. What is the designated disease of cotton?

(A) Black arm
(B) Ashy stem blight
(C) Downy mildew
(D) Red rot

26. Which disease is the designated one of greengram?

(A) Black arm
(B) Ashy stem blight
(C) Downy mildew
(D) Red rot

27. ---------------- is the designated disease of sugarcane.

(A) Black arm
(B) Ashy stem blight
(C) Downy mildew
(D) Red rot

28. In sunflower, the designated disease is -----

(A) Black arm
(B) Ashy stem blight
(C) Downy mildew
(D) Red rot

29. Which one of the following is the designated disease of okra?

(A) Tobacco mosaic virus
(B) Late blight
(C) Yellow vein mosaic virus
(D) Brown rot

30. ---------------- is the designated disease of potato.

(A) Tobacco mosaic virus
(B) Late blight
(C) Yellow vein mosaic virus
(D) Brown rot

31. Which disease is the designated disease of chilli?

(A) Tobacco mosaic virus
(B) Late blight
(C) Yellow vein mosaic virus
(D) Brown rot

32. In tomato, ------------------- is the designated disease.

(A) Tobacco mosaic virus
(B) Late blight
(C) Yellow vein mosaic virus
(D) Brown rot

33. Which one of the following is the designated disease of cauliflower?

(A) Crown rot (B) Mosaic
(C) Black rot (D) Black leg

34. In petunia, -------------- is the designated disease.

(A) Crown rot (B) Mosaic
(C) Black rot (D) Black leg

35. ----------------- is the designated disease of lettuce.

(A) Crown rot
(B) Mosaic
(C) Black rot
(D) Black leg

36. Which disease is the designated disease of cabbage?

(A) Crown rot
(B) Mosaic
(C) Black rot
(D) Black leg

37. Quarantine is

(A) The name of a plant
(B) The name of a character
(C) The process of observation of alien material
(D) None of the above

38. Rice varieties IR 20, IR 26 and IR 36 have shown better resistance to

(A) Tungro
(B) Bacterial leaf blight
(C) Blast
(D) Brown plant hopper

39. Thomas Andrew Knight first noted differences among wheat varieties in their resistance to

(A) Smut (B) Bunt
(C) Rust (D) Wilt

40. Aflatoxins are produced by :

(A) Infected plants

(B) Fungi

(C) Infected bacteria

(D) Nematodes

41. In India, plant quarantine work is done by:

(A) NBPGR (B) TNAU

(C) IARI (D) NRRI

ANSWERHEET

1. **A**	2. **C & D**	3. **D**	4. **D**	5. **D**	6. **D**	7. **A**	8. **B**	9. **C**	10. **B**
11. **D**	12. **C**	13. **A**	14. **B**	15. **C**	16. **A**	17. **D**	18. **A**	19. **B**	20. **C**
21. **D**	22. **B**	23. **A**	24. **C**	25. **A**	26. **B**	27. **D**	28. **C**	29. **C**	30. **D**
31. **B**	32. **A**	33. **D**	34. **A**	35. **B**	36. **C**	37. **C**	38. **C**	39. **C**	40. **B**
41. **A**									

Chapter 2

SEED HEALTH TESTING

1. Which one of the following is the storage fungi?

   (A) Aspergillus (B) Verticillium

   (C) Xanthomonas (D) Ustilago

2. Which one of the following is not physical abnormalities of seed due to storage fungi infection?

   (A) Shrivelling of seed coat

   (B) Reduction in seed size

   (C) Discolouration of seed coat

   (D) Reduction in germination

3. Which one of the following seed moisture content in the field fungi activity is terminated in storage?

   (A) > 10% (B) > 11%

   (C) > 13% (D) > 14%

4. At what seed moisture range, the field fungi activity is excess for their growth?

   (A) 24-25% (B) 23-24%

   (C) 25-26% (D) 26-27%

5. Which one of the following is the field fungi that will also infect on seed storage?

   (A) Aspergillus (B) Verticillium

   (C) Xanthomonas (D) Ustilago

6. At what RH, the storage fungi grow on seed during storage?

   (A) 60-80% (B) 80-100%

   (C) 70-90% (D) 90-100%

7. How many groups of Aspergillus are representing the storage fungi?

   (A) 8-15 groups (B) 9-15 groups

   (C) 10-15 groups (D) 11-15 groups

8. Which one of the following is not relevant for objectives of seed health testing?

   (A) Testing for seed vigour

   (B) Testing for quarantine

   (C) Testing for advisability of seed treatment

   (D) Testing for resistance of cultivars

9. Which one of the following is not a fruiting structures of fungi that can be observed in examination of dry seed?

   (A) Acervuli (B) Mycelium

   (C) Sclerotia (D) Pycnidia

10. For blotter method of seed health testing, how many days the seeds are incubated?

    (A) 4 days (B) 5 days

    (C) 6 days (D) 7 days

11. What is the duration of alternating light and dark that is given for blotter method of seed health testing?

    (A) 6 h (B) 12 h

    (C) 18 h (D) 24 h

12. In agar method of seed health testing, the plated seeds are incubated for ---------------- days.

    (A) 5-7 (B) 4-6

    (C) 3-5 (D) 2-4

13. At what temperature range, the placed seeds are incubated for testing the health status of seed.

(A) 19-25°C (B) 20-25°C
(C) 21-25°C (D) 22-25°C

14. Which one of the following is not a seedling symptom by storage fungi?

(A) Making seeds ungerminable
(B) Rotting of fruits
(C) Rotting of seeds and seedling
(D) Killing of seedling

15. Washing method is not suitable for health testing of which one of the following fungi?

(A) Smut of sorghum
(B) Karnal bunt of wheat
(C) Loose smut of wheat
(D) Downy mildew of soybean

16. ------------% NaOH is used in NaOH seed soak method for testing of storage fungi.

(A) 0.1% (B) 0.2%
(C) 0.5% (D) 1.0%

17. Which one of the following method is suggested for testing of Karnal bunt of wheat?

(A) NaOH seed soak method
(B) Growing on test
(C) Indicator test
(D) Serological technique

18. The Karnal bunt infected seeds become ---- in colour and that can be detected easily.

(A) White (B) Brown
(C) Black (D) Red

19. Which one of the following can be used for confirming the presence of Karnal bunt?

(A) Teliospore (B) Sclerotia
(C) Pycnidia (D) Conidia

20. PCR is used to detect ---------------- of microorganisms.

(A) DNA (B) RNA
(C) mRNA (D) tRNA

21. Growing on test is adopted to test the presence of -------------------.

(A) Fungi (B) Virus
(C) Nematode (D) Bacteria

22. Which method is followed for testing of Xanthomonas?

(A) NaOH seed soak test
(B) Growing on test
(C) Washing test
(D) Blotter method

ANSWER SHEET

1. **A**	2. **D**	3. **D**	4. **A**	5. **B**	6. **C**	7. **C**	8. **A**	9. **B**	10. **D**
11. **B**	12. **A**	13. **D**	14. **B**	15. **C**	16. **B**	17. **A**	18. **C**	19. **A**	20. **A**
21. **D**	22. **B**								

Chapter 3

SEED STORAGE INSECTS

1. What is the storage loss due to insects?
(A) 36-43% (B) 35-42%
(C) 34-41% (D) 33-40%

2. Which one of the following is the percentage loss by rodents?
(A) 36-43% (B) 35-72%
(C) 30-60% (D) 20-50%

3. Loss of seed due to birds is ----------------.
(A) 10-24% (B) 5-15%
(C) 12-34% (D) 22-42%

4. Which one of the following insect damages maize by making a characteristic round holes on emergence?
(A) *Callosobruchus maculatus*
(B) *Corcyra cephalonica*
(C) *Trogoderma granaria*
(D) *Sitophilus oryzae*

5. Which one of the following insect damages pulses by making round holes?
(A) *Callosobruchus maculatus*
(B) *Corcyra cephalonica*
(C) *Trogoderma granaria*
(D) *Sitophilus oryzae*

6. Which one of the following is a scientific name for rice weevil?
(A) *Callosobruchus maculatus*
(B) *Corcyra cephalonica*
(C) *Trogoderma granaria*
(D) *Sitophilus oryzae*

7. Which one of the following is a scientific name for rice moth?
(A) *Callosobruchus maculatus*
(B) *Corcyra cephalonica*
(C) *Trogoderma granaria*
(D) *Sitophilus oryzae*

8. The scientific name for pulse beetle is
(A) *Callosobruchus maculatus*
(B) *Corcyra cephalonica*
(C) *Trogoderma granaria*
(D) *Sitophilus oryzae*

9. Which one of the following crop seed is damaged by rice moth besides rice and cereals?
(A) Redgram (B) Bengalgram
(C) Groundnut (D) Sesamum

10. Which one of the following storage pest can infest the seed in the field before harvest?
(A) Pulse beetle
(B) Rice weevil
(C) Larger grain borer
(D) Confused flour beetle

11. Which one of the following is a common name of *Tribolium castaneum*
(A) Khapra beetle
(B) Red flour beetle
(C) Saw-toothed grain beetle
(D) Confused flour beetle

12. What is the common name for *Oryzaephilus surinamensis*

(A) Khapra beetle

(B) Red flour beetle

(C) Saw-toothed grain beetle

(D) Confused flour beetle

13. The common name for *Tribolium castaneum*

(A) Khapra beetle

(B) Red flour beetle

(C) Saw-toothed grain beetle

(D) Confused flour beetle

14. Which one of the following common name for *Tribolium surinamens* is

(A) Khapra beetle

(B) Red flour beetle

(C) Saw-toothed grain beetle

(D) Confused flour beetle

ANSWER SHEET

1. **A**	2. **B**	3. **C**	4. **D**	5. **A**	6. **D**	7. **B**	8. **A**	9. **C**	10. **A**
11. **B**	12. **C**	13. **A**	14. **D**						

Unit 7
SEED INDUSTRY DEVELOPMENT AND MARKETING

Chapter 1

SEED INDUSTRY

1. In which year, the Royal Commission on Agriculture was constituted?

(A) 1920 (B) 1925
(C) 1930 (D) 1935

2. Which one of the following was established by GOI and state governments on recommendation of Royal Commission on Agriculture?

(A) Research Institutes
(B) Seed Law Enforcement
(C) Seed Testing Laboratories
(D) Seed Certification

3. In which year, the private vegetable seed companies had developed seed production facilities?

(A) 1935 (B) 1940
(C) 1945 (D) 1950

4. The private vegetable seed companies had started seed production facilities in --------- during 1945.

(A) Jammu (B) Srinagar
(C) Simla (D) Quetta

5. In --------------, All India Seed Growers, Merchants and Nursery Men's Association was organised.

(A) 1945 (B) 1946
(C) 1947 (D) 1948

6. In which year, All India Co-ordinated Maize Improvement Program was started?

(A) 1937 (B) 1947
(C) 1957 (D) 1967

7. First maize hybrid was released in the year ----------------.

(A) 1941 (B) 1951
(C) 1961 (D) 1971

8. In which year, All India Co-ordinated Bajra and Sorghum Improvement Program was launched?

(A) 1950 (B) 1960
(C) 1970 (D) 1980

9. First bajra hybrid was released in the year - ---------------.

(A) 1955 (B) 1965
(C) 1975 (D) 1985

10. In which year, the first sorghum hybrid was released for cultivation?

(A) 1984 (B) 1974
(C) 1964 (D) 1954

11. All India Co-ordinated Maize Improvement Program was launched in ---------------- five year plan.

(A) First (B) Second
(C) Third (D) Fourth

12. In which year, the second five year plan was operated?

(A) 1951-56 (B) 1956-61
(C) 1950-51 (D) 1945-50

13. Seed Multiplication Team Review was launched in the year ---------------.

(A) 1961 (B) 1951
(C) 1941 (D) 1971

14. When was NSC established?

(A) 1953 (B) 1963
(C) 1973 (D) 1983

15. Seed Act was launched in the year ----------

(A) 1946 (B) 1956
(C) 1966 (D) 1976

16. In which year, the Seed Rules was implemented?

(A) 1969 (B) 1968
(C) 1967 (D) 1966

17. In which year, the Seed Review Team was launched?

(A) 1969 (B) 1968
(C) 1967 (D) 1966

18. Tarai Development Corporation was established in the year -------------.

(A) 1966 (B) 1967
(C) 1968 (D) 1969

19. When was ISST established?

(A) 1974 (B) 1973
(C) 1972 (D) 1971

20. In which five year plan, TDC was established?

(A) First (B) Second
(C) Third (D) Fourth

21. Seed Act was established during --------------------

(A) First Five Year Plan – Annual Plan
(B) Second Five Year Plan – Annual Plan
(C) Third Five Year Plan – Annual Plan
(D) Fourth Five Year Plan – Annual Plan

22. In which year, first phase of National Seeds Programme was launched?

(A) 1973-74 (B) 1974-75
(C) 1975-76 (D) 1976-77

23. Which state was benefitted by NSP first phase?

(A) Karnataka (B) Rajasthan
(C) Tamilnadu (D) Punjab

24. Which one of the following state was benefitted by the second phase of NSP?

(A) Karnataka (B) Tamilnadu
(C) Punjab (D) Haryana

25. In which five year plan, the first phase of NSP was established?

(A) Fifth (B) Sixth
(C) Seventh (D) Eighth

26. In which five year plan, Central Seed Certification Board was established?

(A) Fifth (B) Sixth
(C) Seventh (D) Eighth

27. IARI was established in the year ------------

(A) 1902 (B) 1903
(C) 1904 (D) 1905

28. When Association of Official Seed Analysts was started?

(A) 1908 (B) 1907
(C) 1906 (D) 1905

29. ISTA was launched in the year -------------.

(A) 1904 (B) 1914
(C) 1924 (D) 1934

30. When ICAR was established?

(A) 1928 (B) 1929
(C) 1938 (D) 1939

31. In which year SFCI (State Farm Corporation of India) was framed?

(A) 1969 (B) 1968
(C) 1967 (D) 1966

32. In which year, the first cotton hybrid was released?

(A) 1969 (B) 1970
(C) 1971 (D) 1972

33. When was Seed Control Order implemented?

(A) 1963 (B) 1973
(C) 1983 (D) 1993

34. New Seed Policy was launched in the year ----------------.

(A) 1988 (B) 1978
(C) 1968 (D) 1958

35. When National Seed Policy was established?

(A) 2004 (B) 2003
(C) 2002 (D) 2001

36. PPV & FR was started in the year ---------

(A) 2004 (B) 2003
(C) 2002 (D) 2001

37. In which year, New Seed Bill was passed?

(A) 2004 (B) 2003
(C) 2002 (D) 2001

38. International Maize and Wheat Improvement Centre (CIMMYT) has its headquarter in :

(A) Switzerland (B) France
(C) USA (D) Mexico

39. ICRISAT deals with

(A) Cotton (B) Maize
(C) Millets (D) Wheat

40. International Rice Research Institute (IRRI) is situated in

(A) Philippines (B) Japan
(C) China (D) Burma

41. The first private seed came into existence in the year

(A) 1918 (B) 1987
(C) 1959 (D) 1912

42. First private seed company was

(A) Monsanto (B) Namdhari
(C) Sutton & Sons (D) Takii

43. The first symbolic importance of seed started after the report of Famine Commission in :

(A) 1881 (B) 1890
(C) 1891 (D) 1892

ANSWER SHEET

1. **B**	2. **A**	3. **C**	4. **D**	5. **B**	6. **C**	7. **C**	8. **B**	9. **B**	10. **C**
11. **B**	12. **B**	13. **A**	14. **B**	15. **C**	16. **A**	17. **B**	18. **D**	19. **D**	20. **D**
21. **C**	22. **C**	23. **D**	24. **A**	25. **A**	26. **D**	27. **D**	28. **A**	29. **C**	30. **B**
31. **A**	32. **C**	33. **C**	34. **A**	35. **C**	36. **D**	37. **A**	38. **D**	39. **C**	40. **A**
41. **D**	42. **C**	43. **A**							

Chapter 2

SEED MARKETING

1. Which one of the following is not pertaining to seed marketing?
   (A) Demand forecast
   (B) Production structure
   (C) Marketing structure
   (D) Sales promotional activities

2. Which one of the following organisation follows single window system of marketing?
   (A) Private seed producer
   (B) National Seeds Corporation
   (C) Co-operative sector
   (D) Agriculture Department

3. Which organisation does both production and distribution of seeds?
   (A) Private seed producer
   (B) National Seeds Corporation
   (C) Co-operative sector
   (D) Agriculture Department

4. Give an example for multiple window system of marketing?
   (A) NSC
   (B) Agriculture Department
   (C) Horticulture Department
   (D) Agriculture Marketing Department

5. Which one of the following is not related to phytosanitary certification?
   (A) Issued for import of seed from other countries
   (B) Issued by phytosanitary official
   (C) Issued for seed purchased within the country
   (D) To check the presence of pests and diseases

6. Which of the following materials are not issued with phytosanitary certificate
   (A) Tissue culture plants
   (B) Sexually propagated material
   (C) Pest attacked seed
   (D) Diseased seed

ANSWER SHEET

1. **B** 2. **A** 3. **C** 4. **D** 5. **B** 6. **C**

Chapter 3

PROTECTION OF PLANT VARIETIES AND FARMERS RIGHTS ACT

1. In which year, PPV & FR Act was implemented in India?

(A) 1999 (B) 2000
(C) 2001 (D) 2002

2. In the year ----------------, the regulations of PPV & FR Act were formulated.

(A) 2000 (B) 2001
(C) 2002 (D) 2003

3. When PPV & FR Authority came into force?

(A) November 2005
(B) October 2005
(C) December 2005
(D) September 2005

4. How many chapters are there in PPV & FR Act?

(A) 9 (B) 11
(C) 10 (D) 12

5. How many sections are provided in PPV & FR Act?

(A) 97 (B) 98
(C) 99 (D) 100

6. Which one of the following is the nodal agency for PPV & FR?

(A) Ministry of Commerce
(B) Ministry of External Affairs
(C) Ministry of Agriculture
(D) Ministry of Co-operation

7. Which one of the following is not related to notable features of PPV & FR?

(A) Protection of plant varieties
(B) Protection of farmers right
(C) Promotion of competitive plant breeding
(D) Ensures supply of good quality seed

8. Which one of the following is not registered with plant variety registry?

(A) Microbes (B) Mangrooves
(C) Mulberry (D) Kusum

9. How many plant species have been registered from May 27, 2007?

(A) 10 (B) 11
(C) 12 (D) 13

10. Which one of the following is not pertaining to the criteria for protection / registration of varieties?

(A) Notable (B) Distinctiveness
(C) Uniformity (D) Stability

11. Which one of the following variety is the derivative of 'Initial Variety'?

(A) Folk variety
(B) Essentially derived variety
(C) Extant variety
(D) Farmers variety

12. Under PPV & FR Act, compulsory license is issued to the registered variety after expiry of ---------- years from the date of registration?

(A) 1 (B) 2
(C) 3 (D) 4

13. How many issues per year the Plant Variety Journal of India are published?

(A) 2 (B) 4
(C) 6 (D) 12

ANSWER SHEET

1. **C**	2. **D**	3. **A**	4. **B**	5. **A**	6. **C**	7. **D**	8. **A**	9. **C**	10. **A**
11. **B**	12. **C**	13. **B**							

Chapter 4

DUS TEST

1. To establish the distinctness of variety, the test is performed for ------------ years for self pollinated crops.

   (A) 1 (B) 2
   (C) 3 (D) 4

2. What is the tolerance limit of offtypes for self pollinated crops while testing uniformity of variety?

   (A) 1% (B) 2%
   (C) 3% (D) 4%

3. While testing uniformity of variety of cereals, how many variants are permitted?

   (A) 1 variant in 150 rows
   (B) 2 variant in 150 rows
   (C) 3 variant in 150 rows
   (D) 4 variant in 150 rows

4. Which one of the following variant limit is considered unstable for cereal while testing stability of variety?

   (A) 8 variants in 150 ear to row progeny
   (B) More than 8 variants in 150 ear to row progeny
   (C) Less than 8 variants in 150 ear to row progeny
   (D) Not less than 8 variants in 150 ear to row progeny

5. Contamination permitted in maize is

   (A) 1% (B) 2%
   (C) 0.10% (D) 0.20%

ANSWER SHEET

1. **B** 2. **A** 3. **C** 4. **B** 5. **A**